IMAGES
of Aviation

BARBERS POINT NAS

On the Cover: Pictured here in the early 1960s is a Kaman HH-43 Huskie helicopter taking off in front of the control tower. (Naval Air Museum Barbers Point.)

IMAGES of Aviation

BARBERS POINT NAS

Brad Sekigawa and Brad Hayes

ARCADIA PUBLISHING

ISBN 978-1-4671-3329-6

Published by Arcadia Publishing
Charleston, South Carolina

Printed in the United States of America

Library of Congress Control Number: 2015931259

For all general information, please contact Arcadia Publishing:
Telephone 843-853-2070
Fax 843-853-0044
E-mail sales@arcadiapublishing.com
For customer service and orders:
Toll-Free 1-888-313-2665

Visit us on the Internet at www.arcadiapublishing.com

Dedicated to the men and women who serve our country.

Contents

ACKNOWLEDGMENTS

First and foremost we would like to thank the individuals who took their time to help us select photographs to tell the story of Barbers Point NAS. We have so much gratitude for the military men and women who helped shape the history of this former naval air station. Images in this book come from a variety of sources—from personal collections of veterans to archived images from the National Archives and Records Administration (NARA), the US Navy (USN), the VW-1 All Hands Alumni Association (VW-1 Association), World War II Database, and of course, the Naval Air Museum of Barbers Point. Without the records and contributions from these organizations, the story of Barbers Point would go untold.

INTRODUCTION

For many local people living on the island of Oahu, and especially to all the military personnel who served there, it will always be known as Barbers Point Naval Air Station. Today it is called Kalaeloa Airport (also John Rodgers Field or JRF), and those who are not versed in Hawaiian culture and history may be surprised to learn that Kalaeloa was the original name of the place. Located on the southern shore of the island of Oahu, and about seven miles west of the entrance to Pearl Harbor, the former naval air station is nestled near the shoreline in the Ewa Plain. The name change occurred more than 200 years ago because of an incident at sea that forever linked the man, who many at the time felt was responsible for a tragic event that could have been avoided if he had only heeded the advice of others, to the place.

Barbers Point was named for Henry Barber, a British sea captain and master of the *Arthur*, a 100-foot two-masted sailing ship known as a brig. The Barbers Point tale first entered Pacific lore when a vicious tropical depression battered the southwest coastline of Oahu in the early evening hours of October 31, 1796. Captain Barber and his crew of 22 men made the *Arthur* ready for their voyage to China. While other captains held their ships in port, Captain Barber, determined to get underway despite the storm, hoisted anchor and departed at 6:00 p.m.

Around 8:00 p.m., the *Arthur*, after being pounded by wind, rain, and heavy surf, struck a coral shoal to the west of Pearl Harbor. Even after clearing the shoal, the British brig was continuously battered by the unrelenting weather and was finally driven upon the reef, where she started breaking up. Captain Barber then gave the order to abandon ship, and the crew took to the lifeboats. Of the 22 crewmen, six drowned in the choppy waters. The survivors and their captain struggled ashore near a tract of land referred to by native Hawaiians as Kalaeloa, meaning long cape or headland. In Hawaiian folklore, it was known as the legendary birthplace and burial ground of Hawaiian kings, but ever since that fateful event, the place would be known as Barbers Point.

Over time, the story of that tragic event was largely forgotten, although Barber's name would be indelibly linked to that desolate part of the island. As the population grew, along with the influx of foreign influence, development in the outlying areas of Honolulu began in earnest, starting with agriculture; in particular, sugarcane. One of the early pioneers of the thriving sugar industry in the Kingdom of Hawaii was James Campbell (1826–1900), who later became one the wealthiest landowners in Hawaii. In the 1860s, as Campbell's business in the sugar industry grew, so did his acquisition of land. Around 1875, Campbell purchased 41,000 acres of a flat, barren area in the Ewa District of Oahu. Criticized for what seemed like a wasteful investment by other businessmen, Campbell nevertheless found artesian wells to irrigate his sugarcane fields and turned arid wastelands into profitable and productive agricultural districts for many years to come. (It should be noted that although James Campbell owned the land in the Ewa Plain, he did not have any sugar mills there. He leased 40,000 acres to Oahu Sugar, a conglomerate of businessmen who built the sugar mills in Ewa and Waipahu.) Even today, the James Campbell Estate still wields influence in the Ewa District and throughout the Hawaiian Islands.

After the overthrow of the Hawaiian monarchy in 1893, another entity would wield even greater influence in Hawaii—the United States of America. The beginning of the 20th century ushered in a new era of global politics and change in the Pacific, with the just-annexed Territory of Hawaii playing a pivotal role. Because of its location, the politicians and military strategists in Washington, DC, saw the Hawaiian Islands as the Gateway to the Pacific, an important hub of commerce that would need protection from other foreign nations. So began the build-up of American military presence in the islands—starting with Pearl Harbor. Later on, with the arrival of the airplane, airfields and bases were needed as well to supplement the expanding military capability of the US Army and Navy. Commercial or civilian air travel was limited to only the outer islands of Hawaii, Maui, and Kauai.

In the early 1920s, the Navy leased 150 acres in the Ewa District from the James Campbell estate for the purpose of erecting a mooring mast for large airships known as dirigibles. In 1925, a 100-foot-tall mooring mast was built on an area of about 3,000 square feet. Besides the mooring mast, other supporting infrastructure was required, including a 10,000-gallon storage tank for aviation gasoline, machinery, housing for mooring the airship, a two-inch water pipe from the nearby Ewa Sugar Plantation, and telephone lines; a railroad track encircling the site was also laid down. Since the mooring mast was built near the Oahu Railway line, it was a prominent landmark for train dispatchers and longtime residents of the area. At the time, the US Navy invested in and explored the use of lighter-than-air balloons and airships culminating in large, rigid-frame dirigibles like the USS *Akron* and *Macon*. Based on, or inspired by, the German zeppelins of World War I, the Navy versions went even further by being able to accommodate as well as launch and recover small biplane fighters known as the Curtiss F9C Sparrowhawk. That was the intent in erecting Ewa Mooring Mast—to have the USS *Akron*, or others like her, be able to moor there—but as it turned out, none of the large Navy airships ever cruised over the Pacific Ocean to Hawaii. The Navy eventually abandoned the large airship concept after several tragic losses, and by 1932, the tall mooring mast was cut down to a shorter stub-mast until its final dismantling in 1942. Near the mooring mast, the Navy also began building a 1,500-foot auxiliary, or outlying field (OLF), in the early 1930s.

Around 1939–1940, the lease to the tract of land had expired, and negotiations began for the acquisition of 7,000 acres of land surrounding Ewa Mooring Mast. The intent was to have the land divided into an emergency landing field and another part turned into a bombing range. This never transpired, but 3,500 acres were eventually acquired. With the acquisition of this new parcel of land came the enlargement of the OLF, which became Marine Corps Air Station (MCAS) Ewa. The site was chosen for its ideal peacetime air-training atmosphere and was completed in early 1941. Ewa Field, as it was known then, was adjacent to the plantation housing for the nearby Ewa sugar mill. The early history of Barbers Point NAS is tied to Ewa Mooring Mast/MCAS Ewa Field because of their close proximity to each other and the origins of the land.

On July 12, 1940, an order was received from Pacific Naval Air Bases (contractors) to survey the newly acquired parcel of land for a freshwater source, as there were known artesian wells where the sugar mills had been established. Topographical surveys and aerial photographs were taken of the land from the time of the order until June 1941. By October 1941, preliminary plans were issued for the construction of Naval Air Station Barbers Point for an estimated cost of $16,400,000. Initial clearing of the land began in November 1941, and the original layout for Barbers Point was prepared by the design section of the 14th Naval District Public Works Department. Initial construction was no easy task, as the entire area was covered with dense, thorny keawe bushes and trees similar to the mesquite found in the Southwestern United States. In areas where there was no vegetation, rough coral hardscape lay with occasional holes and caverns "large enough to fit a house or freight car," according to the *History of Marine Corps Air Station Ewa, O'ahu.* Often, machetes, picks, and shovels gave way to dynamite just to clear the area for surveying, and if the hard ground and thick thorny brush were not enough, the heat in the late summer months made the back-breaking work almost unbearable.

Base construction at Barbers Point was well underway when the Japanese attacked Pearl Harbor on December 7, 1941, but no structures had been built and there were not any aircraft parked at the site. Besides Pearl Harbor itself, other target areas hit by the Japanese were Hickam Field, Wheeler Field, Schofield Barracks, Kaneohe Naval Air Station, Bellows Field, and Ewa MCAS, which is right next to where the new air station was being constructed. Although much of the attack was concentrated on Pearl Harbor, all other targeted installations sustained a great deal of damage and casualties, both military and civilian. Thus, the island of Oahu became the first battleground of the Pacific War and ushered in America's entry into World War II. The Gateway to the Pacific had been severely damaged, and with it loomed the fear of invasion of the Hawaiian Islands by the Empire of Japan. Despite such uncertainty, the recovery and rebuilding of Pearl Harbor and all affected bases would be nothing short of miraculous considering much of it was accomplished within a very short time. Now that war had come to America's doorstep, there was no time to waste—the United States and its allies were facing an unrelenting and brutal enemy that threatened to conquer the entire Pacific region. The war in the Pacific would be bitterly fought at a very high cost.

Since the beginning of the 20th century, the US Navy had been steadily building its fleet of warships as the nation grew. The backbone of the Navy in the years following World War I was the battleship, and America amassed an impressive, if not mighty, fleet of super dreadnoughts throughout the 1920s and early 1930s. The airplane at this time was still relatively new technology and was not seen as a serious threat by the ruling admirals of the world's navies, even with advances in aeronautical design and capability accomplished with each passing year. Only a few young officers in certain navies saw that the battleship's dominance of the seas would be diminished and overruled by the airplane; naval air power would be the deciding factor in any future war waged on the high seas. As the US Navy's fleet of battleships lay crippled in the shallows of Pearl Harbor, the aircraft carrier assumed the spotlight for the war to come, and only a handful of flattops faced the insurmountable task of taking the fight to the enemy. As the Navy began its transformation from a battleship navy to one centered on naval air power, Ford Island at Pearl Harbor and Barbers Point Naval Air Station were key to winning the war in the Pacific, for they were critical in supporting and servicing carrier-borne aircraft.

On April 15, 1942, Naval Air Station Barbers Point was commissioned with 14 officers and 242 enlisted personnel. From then on, the new air station became the center of aviation activity as the Navy strengthened its fleet to carry the war across the Pacific. As the war progressed, the new air station grew rapidly. Standing prominently were the two large hangars, 110 and 111, and the control tower next to them. Within a year of its commissioning, more structures were added including machine shops, an engine-overhaul facility, storage buildings, and accommodations for 2,000 enlisted men, 250 officers, and 800 civilian workers. The civilian workforce proved invaluable for the upkeep and expansion of the air station throughout the war. Initially, Barbers Point could only support two carrier air groups, totaling about 160 aircraft. A typical carrier air group operated three types of aircraft: a single-seat fighter type, a two-seat dive bomber, and a three-man torpedo bomber. On board a carrier there would be two squadrons of a dive-bomber-type aircraft: one bombing unit and one scouting unit, so there would be four squadrons attached to a carrier. The total number of aircraft on board ranged from 80 to 90, depending on the class of carrier. By 1944, as many as four carrier air groups could be serviced at one time at Barbers Point as well as patrol bombers and various types of transports and utility aircraft. The Hawaiian Islands served as the training ground for these carrier groups with the little island of Kahoolawe off the south shore of Maui serving as a bombing target.

Essential to servicing the aircraft of the carrier air groups were the Carrier Air Service Units (CASU), and there were two of these based in Hawaii. CASU-1 was stationed on Ford Island at Pearl Harbor, and CASU-2 was based at Barbers Point. Throughout the war, these two units primarily serviced, prepped, and repaired every type of airplane that flew off the carriers; they were vitally important to the war effort though largely unknown and nearly forgotten. Truly they were one of the many unsung heroes in World War II in the Pacific. From Barbers Point's commissioning until the end of the war, thousands of aircraft of all types passed through Hawaii with one purpose: to defeat the enemy that brought war to its idyllic shores. As the Pacific war ended, many units and squadrons were disestablished as the military transitioned from wartime to peacetime posture, ending that chapter in the air station's history. In 1949, the Coast Guard began air operations at Barbers Point after relocating from Naval Air Station Kaneohe, flying Navy versions of World War II bombers such as the B-24 Liberator (PB-4Y) and the B-17 Flying Fortress (PB-1G). The facility was designated Coast Guard Air Station Barbers Point in 1965 and is the only aviation unit in the 14th Coast Guard District.

The 1950s began with the Cold War escalating in a little known country called Korea, and once again, Barbers Point NAS played an important role in that bloody conflict. It was also the dawn of the jet age. There were no more sporadic encounters as experienced by American fighter pilots battling the German Me-262, the first jet to see combat, but daily aerial duels between Soviet-built MiG-15s and F-86 Sabre jets. Even Marine Corps and Navy jets faced this type of aerial combat, with the Marines scoring the first jet versus jet kill of the war. The Navy embraced this new technology, launching and recovering jet fighter bombers such as the Grumman F9F Panther on World War II carriers designed for propeller-driven aircraft. The learning curve in adapting this new form of carrier operations was very high, but the Navy prevailed. It had only been a short lull since the end of the Pacific War and the beginning of the Korean conflict, and Barbers Point NAS increased air operations by supporting long-range maritime patrol bombers and surveillance aircraft.

The Atomic Age had also arrived, and with it another new threat—long-range Soviet bombers capable of carrying an atomic bomb and submarines with the capability to launch missiles with a nuclear payload. With the end of hostilities in Korea, America and the Soviet Union, along with China, emerged as nations in possession of nuclear weapons incomprehensible in their destructive power. Throughout the 1950s and early 1960s, Barbers Point was home base for long-range submarine hunters and radar-equipped aircraft keeping an ever-watchful eye on the Pacific and Arctic regions for any intruding submarines or bombers crossing over with hostile intent. Based on the Lockheed Constellation airliner, the WV-1, dubbed Willie Victors by their crews, formed the barrier patrols supplementing the Distant Early Warning radar stations that stretched across the Northern Hemisphere known as the DEW Line.

It seemed the Cold War would either end in nuclear holocaust or continue into an everlasting stalemate with occasional tense confrontations or incidents between American and Communist forces bringing both sides to the brink of war. Turmoil erupted in the Far East starting with the French-Indochina War (1946–1954) in another little-known country called Vietnam. America's involvement began in 1955 with military advisors trickling over, escalating into full-blown conflict with the Gulf of Tonkin Incident on August 4–5, 1964. Once again, Barbers Point NAS served as the staging area for naval air power in the Pacific as carrier air wings geared up for the war over Vietnam. At the same time, the Soviet Union increased its submarine force to include more ballistic missile boats prowling the Pacific well within striking distance of the Hawaiian Islands. The twin-engine Lockheed P-2V Neptune, which had served as the primary submarine hunter before the Korean War, was becoming outdated at the end of the 1950s even with upgrades and modifications The takeoff weight of the P2V-7 (SP-2H), the final version of the Neptune, was maxed out for the design of the airframe. It was time for a new antisubmarine warfare platform.

By the early 1960s, a new Lockheed airplane made its debut: the four-engine P-3 Orion, which had its origins as an airliner called the Electra. This airplane, with its long, prominent tail stinger had a longer range, improved submarine-detection sensors, and the ability to carry a greater variety of weapons. The VP Navy, as it was called, had finally matured—a far cry from the early flying boats of the 1920s, through the PBY Catalina of World War II, and the early Cold War P-2V Neptune.

The war in Vietnam finally ended in 1975 with the fall of Saigon, and America exited that disastrous, controversial war that left a nation divided. Barbers Point NAS had already adapted to the changing times in warfare and technology to counter the new threats that were the byproducts of the Cold War. Perhaps the biggest threat lay beneath the ocean's surface: newer classes of Russian submarines prowled the Pacific, and if any came close to Hawaiian waters, the P-3 Orions would depart from Barbers Point in their search for them. For some, it was the ultimate game of hide-and-seek. There was still more saber rattling in the 1980s and 1990s—this time in the Middle East. Although Barbers Point was no longer at the forefront of the world's conflicts, it still played an important role with daily antisubmarine flight operations and its P-3s soldiering onward. When Saddam Hussein invaded Kuwait on August 2, 1990, the very first and only American airplane on the scene was a Barbers Point–based P-3 Orion from VP-1 "Screaming Eagles," deployed to NAS Cubi Point, Philippines. Flying out of NAF Diego Garcia in the Arabian Sea, they began patrolling the Arabian Sea and Persian Gulf, conducting anti-shipping and surveillance missions. As more nations joined the coalition to stop the Iraqi invasion, more P-3 units flew round-the-clock missions, along with Britain and France, in those waters. Hundreds of maritime patrol missions were flown, resulting in 6,300 interdictions of vessels.

When Desert Storm ended, P-3 Orions helped with the sinking of 53 out of 105 Iraqi ships by providing targeting information for carrier air strikes. The end of the first Gulf War also saw the reduction of the US military, not only in the number of personnel and hardware but also bases and facilities. Base Realignment and Closure (BRAC) began, and many communities across the United States that depended on a nearby military base feared the facility would close and cause economic downturn. In 1996, it was confirmed that Barbers Point NAS would indeed close, and the date was set for July 1999. By the mid-1990s, the P-3 Orion squadrons had been reduced from nine to four, and the air station was deemed too large to support that many, which prompted the movement to Kaneohe Marine Corps Air Station (later Marine Corps Base Headquarters-MCBH).

On July 1, 1999, a ceremony was held to officially mark the closure and disestablishment of Barbers Point, the Navy's last naval air station in Hawaii. With more than 3,800 acres and up to 6,500 military and civilian employees and their families, Barbers Point served as the largest naval air station in the Pacific theater. NAS Barbers Point, the Crossroads of the Pacific, was carved out of brush and coral, built amid the ashes of the Pearl Harbor attack, and served through the greatest conflicts of the 20th century. The station that stared down the nuclear stalemate of the Cold War, including Korea and Vietnam, and stood firm against a dictator in the first Gulf War, finally closed after 57 years of service, pride, and tradition.

One

Birth of an Air Station

Construction for Barbers Point began almost immediately following the attack on Pearl Harbor. The emphasis was on Ewa Field, as it was nearing completion of its facilities and runways, which were critical for the air defense of Hawaii while the new air station was under construction. Mobilizing the manpower and equipment was a monumental task considering the devastation at all major military installations; much repair and rebuilding was sorely needed for the coming war.

Nevertheless, Barbers Point NAS was officially commissioned on April 15, 1942. By a conservative estimate, more than 4,500 civilian workers of all trades were hired locally and from the mainland (on one-year contracts) to build the air station. All personnel were carefully screened and their backgrounds checked, especially those local workers of Japanese ancestry whose skills were coveted in such uncertain times.

The photographs in this chapter come from a pictorial record book of all those who built the naval air station in 1942–1943 in some of the harshest conditions. Due to wartime restrictions, very few photographs exist of the actual construction on what was known then as the reservation. Enduring blistering heat, thick thorny brush, biting and stinging insects of all types, and unforgiving terrain, these men worked tirelessly to achieve a near-impossible feat.

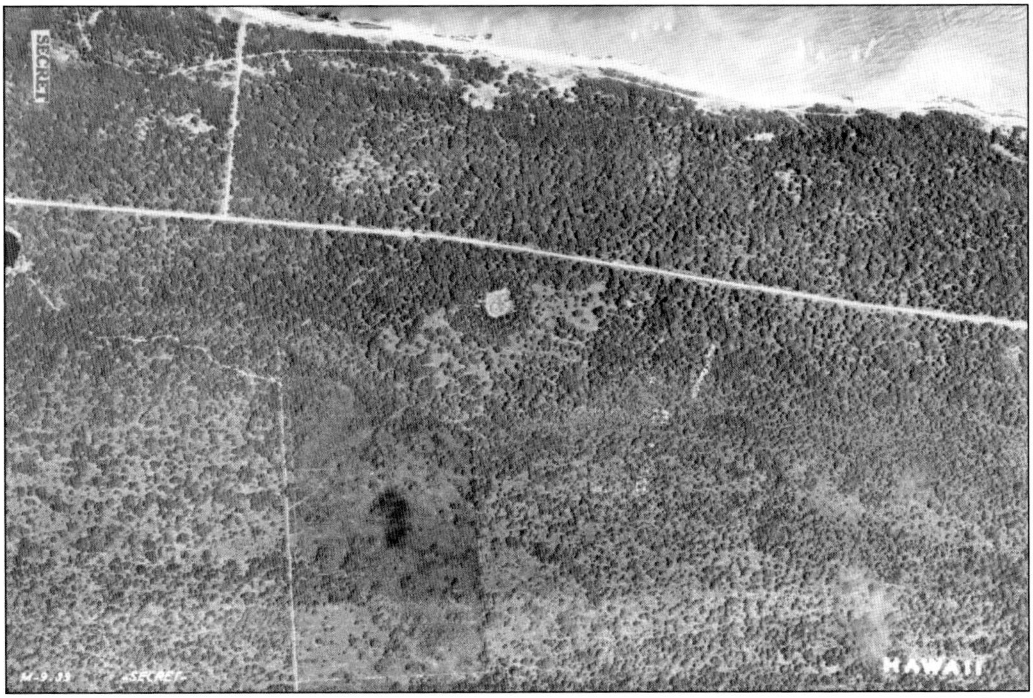

This image taken by the US Army Air Corps 11th Photographic Reconnaissance Squadron from an altitude of 10,000 feet shows the future site of Barbers Point NAS. The main dirt road led to an Army camp while access to the beach and shoreline was by the smaller road, or pathway. The light-colored spot at center below the main road is a 400-foot-diameter survey site. The main runway would be oriented diagonally from lower left to upper right. (Hawaii State Archives.)

Following the attack on Pearl Harbor, defenses around military installations increased throughout the Hawaiian Islands. At Ewa Field, these antiaircraft guns were installed around the perimeter and near the construction site at Barbers Point. This is a five-inch, 25-caliber antiaircraft gun, the same type used on battleships and heavy cruisers. (NARA.)

A US sailor stands guard at another gun site sometime in mid-1942. Wearing bell-bottomed dungarees and a denim shirt, he is armed with a 1903 Springfield bolt-action rifle and a World War I doughboy helmet of British design. Since these were shipboard guns, they would be manned by US Navy personnel. The site itself is still under construction, as they would have a concrete revetment or sandbagged walls surrounding the site. (NARA.)

All the antiaircraft guns protecting Ewa Field and Barbers Point were controlled by these range finders. This is an Mk-33 Gun Fire Control System, the same type used aboard battleships and cruisers. It provided tracking information for the guns if enemy aircraft came within range. The exposed site indicates that construction is not completed yet, as no protective revetment or wall has been set up. Large movable generators were used to power these range finders. (NARA.)

Pictured are the two men who were responsible for overseeing the entire Barbers Point project, A.H. Ayers (above) and Howard A. Schroedel (below). Ayers was the overall project manager, while Schroedel was the general superintendent. Coordinating all the trades and workers was a seemingly insurmountable task given the limited building materials and equipment available at the time as those things were needed elsewhere: namely, Kaneohe Naval Air Station, Wheeler Field, Hickam Field, and Pearl Harbor, where there was much damage to the infrastructure. These two men had the huge burden of responsibility in getting Barbers Point NAS built on time. This project was extremely urgent; Barbers Point had to be operational within one year, as the Navy needed a large air base to conduct all types of air operations. (Both, NAMBP.)

E.S. Mollenhoff, on the far left, was often called "Mollie" and served as the head project engineer for Barbers Point. With his trademark pipe, he made daily tours of the various project sites. (NAMBP.)

These are the surveyors for the Barbers Point project, one of several groups of field engineers. A number of these photographs were taken when the work party arrived on site. They rode a shuttle bus that went to various pickup points outside the base. This practice was continued throughout the project until enough housing was completed for civilian workers. (NAMBP.)

Standing beside their shuttle bus are the timekeepers; all civilian personnel had to clock in just like at any other workplace. Wages for these workers ranged from 75¢ to $2 an hour for unskilled labor. The average civilian male worker earned about $3,600 per year, but women earned considerably less. The average Navy seaman first class and Army private earned around $600 annually. (NAMBP.)

Above is a 65-ton Mack diesel dump truck with an open cab. The roof of the cab was attached to the front of the dumper. There were several dozen of these huge trucks rolling day and night hauling away coral and rock debris when the runways were under construction. Note that on the hood of the truck a US Navy serial number is painted. As this was an official Navy project, all vehicles had to be certified for government use. Below is a photograph of the dump truck drivers. There was a night shift crew as well, but work was limited at night due to blackout conditions. A full-service garage kept all the heavy machinery operating, including mechanics and oilers; hydraulic and other specialists were always on hand. (Both, NAMBP.)

Pictured here are two groups of local Japanese carpenters. These men and others like them were exempt from internment because of their skill and value to the overall war effort. They went through FBI background checks in order to work on any military installation or government project site. It is interesting to note that once the Barbers Point project was finished, many of these young men joined the US Army, forming the famous 442nd Regimental Combat Team. These men most likely ended up in the engineering battalion by virtue of their carpentry skills. (Both, NAMBP.)

The group of men above are the cement finishers and tile setters. They did the fine work following the concrete crew, who poured the base foundations for all the structures. The concrete crew, pictured below, also poured the parking apron, which appears as a white square in aerial photographs. The thought of the Japanese invading the island was a constant concern, especially for the civilian population, who were often kept in the dark about military operations. (Both, NAMBP.)

These are some of the few in-action photographs taken from the book *War Workers in Hawaii* (1942). The building in the background at right is the same structure in the photograph at left. These were the chow hall and community centers for the servicemen on base. The steel girders are being prepped for building the large hangars A and B (now called 110 and 111). (Both, NAMBP.)

The Barbers Point line crew, one of the several groups of electricians who strung up power and telephone lines around the base, is pictured here. This particular group was called the Birchfield Light and Power Company, a father-son outfit. Thomas Birchfield stands at far left, and his son Walter is in the front row kneeling, third from left. Tom Birchfield was there from the very beginning of the project. Also in the group are a mix of mainland and local electricians. (NAMBP.)

Pictured here is the grading and excavation crew. These are just a few of the men who made the runways at Barbers Point as well as all the paved main roads around the air station. In fact, the entire project site had to be graded level before any other work was done. Much of the area was made up of coral, on which picks and shovels could not be used. Dynamite and heavy machinery had to be called in to do the work. (NAMBP.)

A trench line is being dug through the cane fields for water supply pipes off the reservation. (NAMBP.)

Judging by the amount of cooks and bakers, the kitchen must have been very busy with the amount of workers to feed daily. Keep in mind that these meals were prepared before prepackaging, and while everything was made fresh, it was laborious and time consuming. Coordination and preparation were key. The waitresses served the men in the dining hall and the officers and upper management in a separate area. The young Filipina women below have arrived to work in the Barbers Point shuttle bus. (Both, NAMBP.)

SHOPPIN' FOR SMILES: USO chorus girls high light revue for war workers at Area 3.

★

HARMONICA DUO: Ed Shulman and Stan Wisser, former Major Bowes amateurs, who work nights at Pearl Harbor and entertain on USO camp show, "Shoppin' for Smiles."

★

BOSSES ENJOY USO CAMP SHOW.

In the days before television there was just the radio, incomprehensible in this day and age. The only other form of entertainment besides movies was the live show. For the war worker at Barbers Point, this was something to look forward to on any given night. Hula dancers were popular among the workers from the mainland, and there were even talent shows among the workers themselves. The USO played a very important role in keeping spirits up among military and civilians alike. (NAMBP.)

WAR WORKERS
IN HAWAII

At Play
At Barber's Point

★

HAWAIIAN HULA: Especially popular among war workers newly arrived from the mainland are the Hawaiian hula dancers. Here lovely Mabel Lee dances the hula, accompanied by the musical trio, Randy Oness, Pua Almeida, and Buddy Peterson.

AUDIENCE REACTION: Group of workers at Barber's Point shown watching a Wednesday night USO camp show in huge recreation hall 195 feet long by 46 feet wide, which was built for use of the men during their leisure hours.

Many of these men who helped build the air station are largely forgotten or unknown—only the original book printed in 1943 acknowledges their accomplishments or records their even being there. The original photographs are long gone or packed away and forgotten; perhaps they will turn up some day and be rediscovered. (NAMBP.)

This aerial view looking north shows Barbers Point NAS nearing completion in September 1943. Note that the main runway is offset or staggered; this would be enlarged by early 1944. The photograph also shows the close proximity to Ewa Field MCAS. At the same time when construction was underway at Barbers Point, Ewa Field was being improved to include aircraft revetments just off the taxiways. At Barbers Point, the two large hangars can be seen just beyond the concrete apron. (NAMBP.)

Two

THE WAR YEARS

Commissioned on April 15, 1942, construction started in earnest on the new air station, and less than a year later, on April 4, 1943, Barbers Point became operational. During construction throughout 1942, nearby Ewa Field served as the main area for aircraft servicing and maintenance for Navy carrier aircraft. Carrier Aircraft Service Units (CASU), which would service and repair all carrier aircraft heading to the Pacific theater, were critical to the war effort. CASU-2 was commissioned at Ewa Field on May 6, 1942, and remained there until Barbers Point facilities were usable. Just prior to the Battle of Midway (June 4–7, 1942), CASU-2 serviced the air groups from the carriers USS *Hornet* and *Yorktown*.

The Battle of Midway turned the tide in favor of the United States in terms of naval air power in the Pacific. Japan lost four fleet carriers (*Akagi*, *Kaga*, *Soryu*, and *Hiryu*) to the loss of one US carrier, the USS *Yorktown*, but the rest of 1942 would prove to be tenuous for America, as the country lost two more carriers (USS *Wasp* and USS *Hornet*) and have another damaged, the USS *Enterprise*.

It was not until the arrival of new *Essex* class carriers in 1943 that the pace quickened for victory in the Pacific in terms of naval air power. NAS Barbers Point was at the forefront of this effort, as thousands of aircraft and personnel passed through, taking the fight to the enemy's homeland of Japan. By mid-1945, after more than three years of conflict, the end of the Pacific war was in sight.

In this western-facing aerial image from April 5, 1944, Barbers Point NAS now has a larger main runway and is no longer staggered. To the right is Ewa Field, in close proximity to Barbers Point, and both airfields show significant development in the surrounding areas. Just to the left at Ewa Field are aircraft revetments next to the runway. (NARA.)

On September 16, 1943, a US Navy photographer pointed his camera at the parking ramp where part of Air Group 16 resided, having been serviced by CASU-2. SBD-5s, TBF-1 Avengers, and F6F-3 Hellcats can be seen. An SBD-3 Dauntless dive bomber sits near the fueling station, and to the right is a Grumman J2F-3 Duck amphibious floatplane used for utility work. In the far distance smoke belches from an Oahu Railway train used to ferry troops and supplies from Pearl Harbor to the Ewa area. (NARA.)

The photographer shifted his lens to where the J2F is parked and snapped this photograph of the newly completed project 458, also known then as Hangar A (later 110). Just beyond that is Hangar B (111). Three TBF-1 Avenger torpedo bombers are parked; their wings are folded with the unique Grumman design for carrier storage. Next to them is a Douglass SBD Dauntless dive bomber also awaiting servicing by CASU-2. (NARA.)

Moving to the fueling station, the photographer documented the just-completed control tower and operations center also known then as Building 4. Over the years the tower itself went through changes in appearance. (NARA.)

Across the street from Hangar B (111) and the flight line is the newly completed administration building. This would be the base commander's office until Barbers Point closed in 1999. Today, it is the home of American Renaissance Academy, a private school. (NARA.)

Carrier Aircraft Service Unit Two (CASU-2), integral to the operation of Barbers Point, was tasked with every phase of preparing a carrier-borne aircraft for service. Carrier aircraft were a priority, but CASU-2 serviced just about everything the Navy flew in its inventory. Pictured here around 1944 is the metal shop unit responsible for the airframe and the outer skin of aircraft. (USN.)

When this photograph was taken, CASU-2 had reached a milestone—its 1,000th F6F Hellcat repaired. In any air operation, accidents abound; landing, taxiing, or moving aircraft around the field can cause varying degrees of damage from a small dent to whole sections of an airplane shredded or chopped by whirling propellers. Blown engines, electrical problems, fires, and mechanical failures of every sort kept the men of CASU-2 busy throughout the war. Since thousands of the newer combat aircraft were rushed into wartime service, many required tweaking or troubleshooting before being sent to combat duty. (USN.)

On September 1, 1943, the A&R Hangar (assembly and repair) was usably complete. In the photograph, four Grumman F6F Hellcat fighters sit outside. To the right of them is a Curtiss SB2C Helldiver. The building on the left is the power station, which provided electricity to the entire base. (NARA.)

In this remarkable photograph taken from the USS *Pensacola*, a Japanese Type 99 Val dive bomber is on fire and making a suicide plunge into the USS *Hornet* during the Battle of Santa Cruz, October 26, 1942. With the loss of the USS *Wasp* on September 15, 1942, and the USS *Enterprise* damaged in this battle, the USS *Saratoga* would be the only operational carrier left in the Pacific. (World War II Database.)

Pictured here in the Solomon Islands in 1943 is the USS *Robin* (HMS *Victorious*), on loan from Great Britain to help bolster the US Navy's nearly depleted carrier force after losses at Coral Sea (USS *Lexington*), Midway (USS *Yorktown*), and Santa Cruz (USS *Wasp* and USS *Hornet*), and the damaged USS *Enterprise*. On board the USS *Robin* (named for Robin Hood) are US Navy TBF Avengers and F4F-4 Wildcat fighters. (World War II Database.)

HMS *Victorious* (USS *Robin*) on the right and the USS *Saratoga* lie at anchor at Noumea, New Caledonia, in 1943; both carriers had their aircraft serviced by CASU-2 at Barbers Point before departing for operations in the Southwest Pacific. America at this time was rushing to get new aircraft and ships out to turn the tide against the enemy after costly battles in 1942. (Australian War Memorial.)

With the allies firmly on the offensive, Japan and its allies were on the brink of defeat, but even with overwhelming numbers in men and equipment the end was still not within sight. Here, four new *Essex*-class carriers lie at anchor at Ulithi Atoll in the Caroline Islands on December 2, 1944. From left to right are the *Wasp*, *Yorktown*, *Hornet*, and *Hancock*, viewed from the *Ticonderoga* with her deck packed with F6F-5 Hellcats. (World War II Database.)

US Navy pilot Lt. J.W. Stubblebine gets ready to board his TBM-3 Avenger torpedo bomber sometime in late 1944. The classic pose epitomizes the resolve of these men willing to take the fight to the enemy and end the war. Lieutenant Stubblebine is also armed with a personal sidearm and extra ammunition, which was typical for all US airmen in the Pacific. (World War II Database.)

A flight of TBM-1C Avengers of VT-18 from the USS *Intrepid* in the Pacific head out toward Japanese-held territory. Late in the war, the Avenger was rarely used for torpedo-strike missions, as the Japanese navy had been decimated by Allied air strikes and submarine attacks. They could not keep up with the losses, as no new ships were built to replace the ones lost. (World War II Database.)

Pres. Franklin Delano Roosevelt is pictured here on July 27, 1944, touring Barbers Point NAS and Ewa Field MCAS. Also with him are Gen. Douglas MacArthur and Adm. Chester Nimitz, commander in chief of the Pacific Fleet. This was President Roosevelt's second visit to Hawaii; the first was in 1934 when he was newly elected as president and America was at peace. General MacArthur and Admiral Nimitz were apprising FDR of the next big push into the Pacific—the island-hopping campaign and MacArthur's eventual return to the Philippines. (World War II Database.)

Pictured here in 1944 is an overhead aerial view of Barbers Point NAS. Within the white square and on the asphalt parking ramp, more than 300 aircraft are lined up in neat rows, wingtip to wingtip. This would be a common sight throughout 1944 until the end of the war. (USN.)

A North American SNJ two-seat advanced trainer aircraft flies over Honolulu after the war in 1947. Also known as the T-6 Texan for Army Air Force use, the Navy SNJ was virtually the same except later versions had a tailhook for carrier landings. At Barbers Point, these planes were used for maintaining pilot proficiency. Just above the wing in the photograph, the National Cemetery of the Pacific (or Punchbowl Cemetery) can be seen amid the urban sprawl. (USN.)

Pictured here are the control tower and Hangars A and B in early 1945. By this time, trees and some landscaping have taken root at the base of the tower. On the walls of the hangars are warning signs painted in red letters reading, "Beware of Propellers." (NARA.)

An OQ-2 Radioplane drone sits on a launch catapult for use in gunnery target practice at Barbers Point in 1945. These radio-controlled model planes were invented by British actor Reginald Denny, who started an RC model-plane hobby shop in Hollywood between acting jobs. When war broke out, he was contracted to produce model planes for the military to be used as target drones. His company produced more than 15,000 drones during the war. These drones were usually painted a bright red for gunners training on the 20-millimeter Oerlikon autocannon. (NAMPB)

Bagley Amphitheater, where official base-wide announcements, ceremonies, and USO shows were held, was completed in 1944. It was razed during the 1990s and a medical clinic built on the site. (NARA.)

One of the workers at Reginald Denny's company assembles OQ-2 Radioplane drones in a photograph printed in *Yank/ Army Weekly*. Norma Jean Baker would later be known as Marilyn Monroe after being discovered working at that little workshop in Van Nuys, California. A few years later, the famed movie star made a stop at Barbers Point while on her way to entertain troops during the Korean War. (NARA.)

Paul Newman, another future movie star, served in the Navy during the war and was stationed at Barbers Point, undergoing training as an aerial gunner in the TBF/ TBM Avenger before being sent overseas. He could not qualify as a pilot because of his colorblindness. Other Hollywood actors who were also stationed there were Robert Stack (*The Untouchables*), who was a gunnery training officer, and Richard Boone (*Have Gun, Will Travel*), who trained as an aviation ordnance man and gunner on Avengers. (NARA.)

A very lucky sailor gets a kiss from popular singer/entertainer Frances Langford during a USO show in 1945. In the years following the war, many entertainers came through to boost the morale of the troops, including Frank Sinatra, who was filming *From Here to Eternity* in 1953. (NARA.)

On January 30, 1945, the first WAVES (Women Accepted for Volunteer Emergency Service) officer arrived at Barbers Point. WAVES played a vital role in the war effort. While doing mostly clerical work, many held positions in aviation support, communications, medical, intelligence, and science and technology. The only one identified in the photograph is Capt. Mildred H. McAfee, seated at far right, the first director of the WAVES. When this photograph was taken at Barbers Point, she commanded 82,000 women in service. (NARA.)

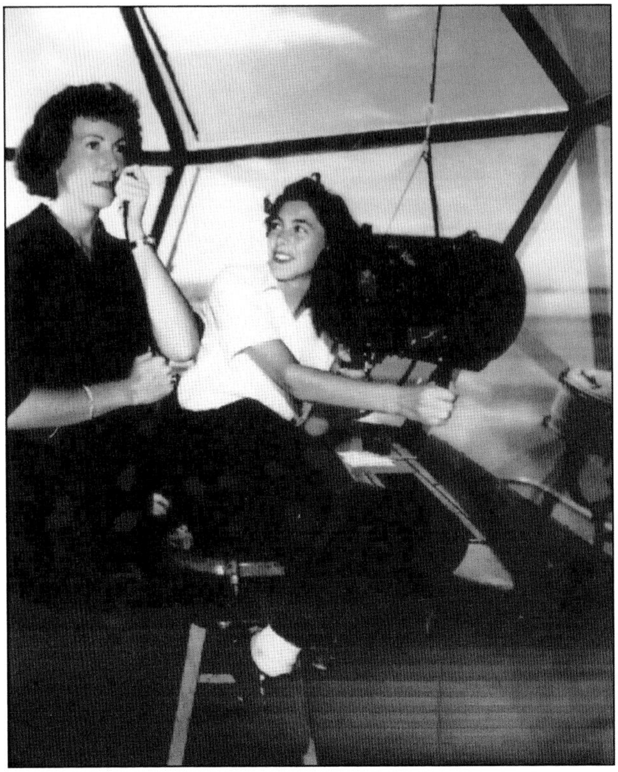

Two WAVES are at the top of the control tower at Barbers Point in 1945. Lois Stoneburg is on the radio, while Mary Johnson mans the light gun for signaling aircraft. Both held the rank of specialist second class. WAVES held the same rank and pay as their male counterparts during World War II. (NARA.)

Photographed in early 1945 is the all–African American Navy band from Barbers Point called the Hellcats, who were considered the best in the Hawaiian Islands. A number of these musicians played professionally in other big bands before enlisting in the Navy. Some of them included Duke Ellington, Les Hite, Ella Fitzgerald, Count Basie, and Fats Waller. The singer is L.K. Oakes, bandmaster and lead musician. (USN.)

Three

The Changing Tide

With the end of World War II, peace had finally returned, but with it came a numbing realization that the world had changed and possibly not for the better. Although the guns had fallen silent, the roar of aircraft continued to reverberate around Barbers Point. Along with the familiar sound of propeller-driven aircraft came a strange sound, the high-pitched whine of a new technology making its debut in the skies of Hawaii—the jet age had arrived.

It was also the dawn of the Cold War, which would last for generations to come. The rumor of war with the Soviet Union and its allies commenced amidst the demobilization and drawdown of the US military. The stakes were high, as Communism threatened to engulf many nations in Eastern Europe and the Far East, escalating in a little-known country called Korea.

As the Cold War heated up in Korea, it saw a new kind of air war—jet versus jet combat—and another type conducted by the US Navy in World War II, night fighting. Barbers Point Utility Squadron One (VU-1) would prove essential to Navy and Marine pilots flying night missions in Korea. At the same time, maritime patrols with long-range land-based Navy bombers flew constant antisubmarine missions as the Soviet Union continued to build its navy.

Following the atomic bombings of Hiroshima and Nagasaki on August 6 and 9 respectively, September 2, 1945, saw the official end to World War II in the Pacific. On the deck of the battleship USS *Missouri*, Gen. Yoshijiro Umezu signs the surrender document as Gen. Richard K. Sutherland witnesses. Gen. Douglas MacArthur stands on the left at the microphone. The USS *Missouri* is now preserved at Pearl Harbor, very near to the USS Arizona Memorial, symbolizing the beginning and end of the war in the Pacific. (NARA.)

Pictured here are morning colors at Barbers Point NAS in 1945. With the Waianae Mountains in the background, members of the Barbers Point band the Hellcats, who also doubled as the official Navy band for important occasions, play the national anthem. (NARA.)

This 1945 aerial view of an Ewa plane pool shows a large collection of Navy and Marine aircraft sitting idle as their wartime service comes to an end. Many of these airplanes ended up in salvage or the scrap-metal heap, while some continued to serve in other capacities. In the foreground lies more wreckage of aircraft and parts no longer needed. (USN.)

At Ewa Field, a row of Curtiss SB2C-4 Helldivers from a Marine squadron sit silently, awaiting disposition for either the scrap dealer or a reserve squadron. Appearing fresh from the factory or with very little flight time, thousands of aircraft of all types ended up in the boneyard on the mainland. Others returning on carriers were simply pushed off the flight decks if they were deemed war weary. (NARA.)

Some airplanes found continued life with other branches of service, as this former Consolidated PB-4Y Privateer attests. Once a Navy patrol bomber, and stripped of its armament, it was modified to fit the needs of the Coast Guard. It was now tasked to fly search-and-rescue missions in Hawaiian waters. Pictured here in March 1949, the Coast Guard air unit is in the process of moving from its former home base at Kaneohe Naval Air Station to its new one at Barbers Point. (NARA.)

FALLWEATRAPAC, painted on the fuselage of this F6F-5N Hellcat, stood for "Fleet All Weather Training Unit Pacific." The unit was established in 1944 as VFAW-3 to train Navy pilots for night missions, and after several designations, became FAWTUPAC (same meaning as above) in 1948. Even after the change, all aircraft retained the original lettering. This airplane is now on display in Michigan at the Kalamazoo Air Zoo. (USN.)

An F6F-5N Grumman Hellcat night fighter in flight off Oahu shows the right-wing mounted radome and 20-millimeter guns. This photograph was taken after 1947, as the national insignia has red stripes, which are barely visible here. The plane was finished in overall glossy sea blue and is missing the national insignia on the top left wing. (USN.)

This overhead view of Barbers Point NAS from 1950 shows the concrete ramp nearly empty. The asphalt apron next to Hangar B (111) is crowded with aircraft parked neatly, most likely belonging to FAWTUPAC. The surrounding area shows continued improvements in base facilities to include more senior officers' housing and a community swimming pool. At the top of the photograph, and along the shoreline, the railroad track can be seen skirting the runway. (USN.)

Taken around the same time as the previous photograph, this is an oblique view of Barbers Point and the deserted Ewa Field. By this time, Marine aviation was moving to Kaneohe, which the Navy had turned over, and by January 15, 1952, recommissioned as a Marine Corps Air Station. On June 18, 1952, Ewa Field was officially closed and turned over to Barbers Point. (USN.)

Rotary-winged aircraft (helicopters) were a new sight at Barbers Point. Here, a Sikorsky HO-3S has just landed during a flight demonstration showcasing their potential capability for search-and-rescue as well as utility work. These aircraft proved invaluable for such tasks during the Korean War. (NAMBP.)

A flight of three Grumman F7F-3N Tigercat night fighters from FAWTUPAC, as their tail codes indicate, fly in formation over the Waianae Mountains. Too late to see action in World War II as a single-seat fighter, a number of these were modified as a two-seat night fighter in the Korean War, with one Marine aircraft claiming the first kill at night on July 1, 1951. The camera lens has stretched the near Tigercat in the photograph, making it appear longer than the others. (USN.)

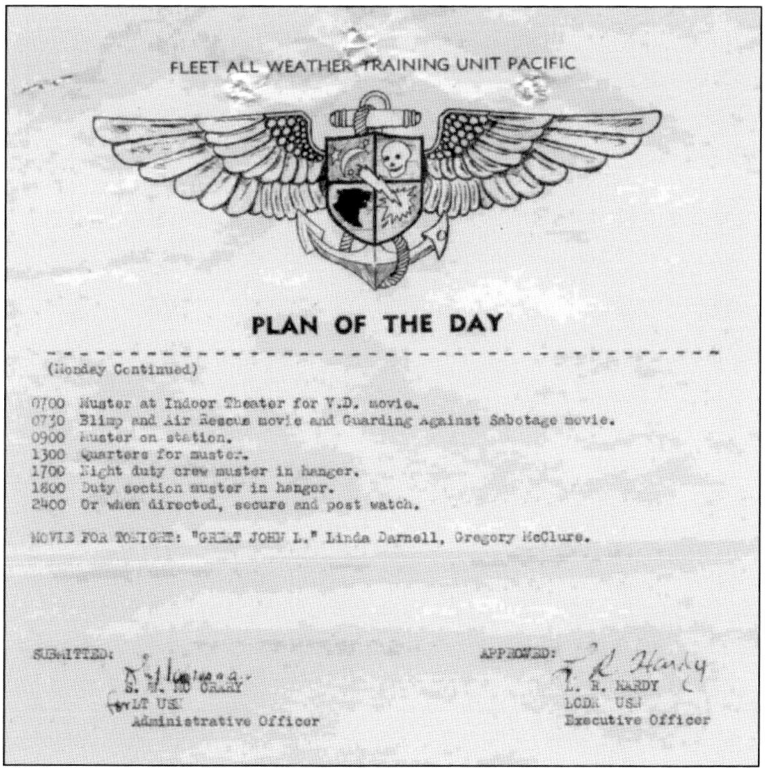

FLEET ALL WEATHER TRAINING UNIT PACIFIC

PLAN OF THE DAY

(Monday Continued)

0700 Muster at Indoor Theater for V.D. movie.
0730 Blimp and Air Rescue movie and Guarding Against Sabotage movie.
0900 Muster on station.
1300 Quarters for muster.
1700 Night duty crew muster in hanger.
1800 Duty section muster in hanger.
2400 Or when directed, secure and post watch.

MOVIE FOR TONIGHT: "GREAT JOHN L." Linda Darnell, Gregory McClure.

SUBMITTED: S. W. MC CRARY, LT USN, Administrative Officer

APPROVED: L. R. HARDY, LCDR USN, Executive Officer

Every squadron and unit was issued a Plan of the Day every morning. Unless one had duty elsewhere, attendance was mandatory, especially for the movie at 7:00 a.m. At least there was a regular Hollywood movie at night. (NAMBP.)

NOTICE: Remember the Bible Study Hour which is held every Sunday evening in the Protestant Chapel at 1900. All hands are invited. Chief Archambe is in charge.

MOVIE FOR TONIGHT: "MR. PEABODY AND THE MERMAID", William Powell, Ann Blythe.

- -

PLAN OF THE DAY — Monday 3 January 1949

CDO LCDR CRAIG
FAD LT KURTZ
SDO LTJG DUCK
AFAD NARDICHIA, A.R.
ASDO SKAGGS, D.D.
BULLER, H.A.
SANDERS, H.K.

DUTY SECTION # 3
SECTION LEADER VESSOL
DUTY DRIVER HOLLINGSWORTH - SUMMERS
DUTY YEOMAN

DUTY MASTER AT ARMS BANKS
BEDMAN
SINCLAIR
PERSONNEL YEOMAN CARTER - COOK - WOOTON

POOW
REID AD2 1300 - 1600
MAXON AD2 1600 - 2000
KEY AD3 2000 - 2400
POWELL AD3 2400 - 0400
CHASE AD3 0400 - 0700

POST # 1
PRICE, C.H. 2300 - 0300
RUSSELL, O.J. 0300 - 0700

POST 3 & 4
GOFF, B.H.
FREEMAN, D.W.
JONES, H.
KRAUS, J.V.
LEE, B.L.

POST # 2
NELSON, G.A.
LANEY, G.G.

BKS. 12: MCCORKORY, M.A.; PITT, E.W.
BKS. 13: BOJIC, D.J.; MORALES, D.
2200 - 0200; 0200 - 0600
BKS. 5: STRATTON, R.G.; SHUMAN, D.R.
BKS. 64: WOOD, W.R.; STEWART, T.A.

In the days before technological gadgets, there was really not much else to do after duty hours were over. Other than reading a book, one could go to a Bible study or watch a movie at the base indoor theater, released the year before. First-run feature films played at movie houses in downtown Honolulu, accessible only on weekend liberty. (NAMBP.)

A Douglas AD-4N Skyraider night fighter of FAWTUPAC is pictured here in July 1952. Based on the original single-seat fighter, the 4N had a crew of three with two radar operators in the fuselage. Crammed with bulky radar equipment, the interior was very small for the two crewmen; the entry door is located next to the national insignia. (USN.)

Flown by the Air Force as the Lockheed P-80 Shooting Star, the Navy version was designated the TV-1; although both were identical, the Air Force version could not land or take off from a carrier deck. A PA 50 sits next to Hangar 110 on July 20, 1952, possibly the only one of its type in FAWTUPAC. (USN.)

This is the official squadron patch of Fleet All Weather Training Unit Pacific (FAWTUPAC) as worn by officers on their leather jackets. The shield was black and red with a white skull; the explosion, lightning bolt, moon, and stars were yellow. (NAMBP.)

A TV-1–coded PA 50 is pictured here in flight off Waikiki on November 12, 1953. The aircraft was finished in a light gray color. The slow arrival of new jet fighters for the Navy prompted the service to acquire a total of 49 aircraft to train new pilots for jets, and at least one example flew out of Barbers Point. On the Waikiki shoreline, only two hotels can be seen, the Royal Hawaiian (left), and the Moana Hotel. (USN.)

Designed as a carrier-based night fighter, the Douglas F3D Skyknight saw action in the Korean War, scoring the first jet-versus-jet kill at night. A USMC F3D-2 of VMF(N)-513 intercepted and shot down a Yak-15 in November 1952. PA 82 is seen flying near the north shore of Oahu in 1953. The Skyknight is painted in the standard paint scheme of overall dark sea blue. (USN.)

A beautiful tight-formation flight of VU-1 aircraft consisting of three Douglas JD-1s (front) and five Grumman TBM-3U Avengers is pictured here. The JD-1 was the same as the Air Force's A/B-26 Invader except these VU-1 aircraft had their guns removed. These planes are painted in a high-visibility paint scheme, as they were used for towing target banners for aerial gunnery practice. The wings and tailplanes were painted yellow with red stripes, and the fuselages were painted seaplane gray. (USN.)

An SNB-1 Kansan, better known as the Beechcraft Model 18, or Twin Beech, joins up with a Douglas JD-1 near a cloud bank. VU-1 had the tail code UA, sometimes called the Unique Antiquers; the squadron was established on July 20, 1952, at Barbers Point, providing training and support for fleet units due to demands of the Korean War. (USN.)

With billowing clouds forming over the Koolau Mountain Range as a backdrop, a trio of VU-1 aircraft head back to Barbers Point in formation. The horizontal and vertical fins were painted yellow, with the rudders painted in red. VU-1 would operate a wide variety of aircraft through the years, including helicopters and jet fighters, prompting a squadron-unit designation change to VC-1 with the "U" in utility becoming "C" for composite. (USN.)

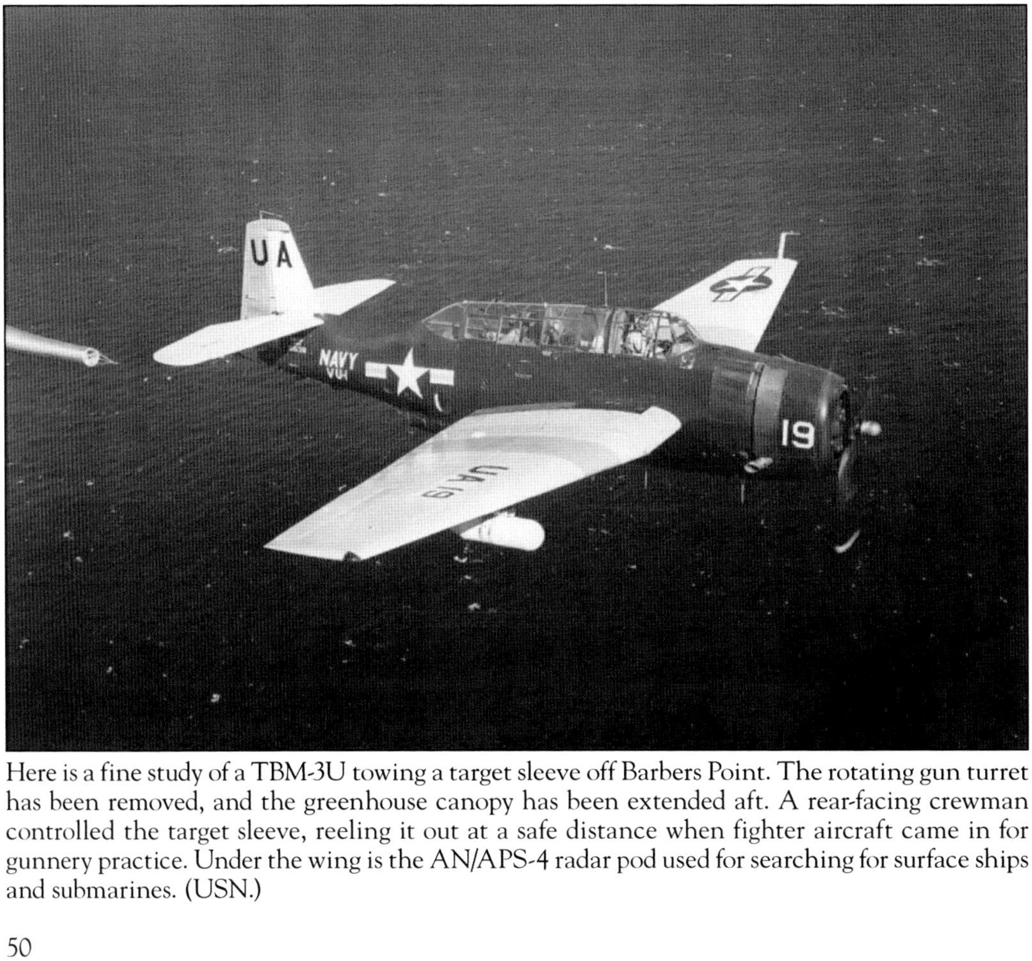

Here is a fine study of a TBM-3U towing a target sleeve off Barbers Point. The rotating gun turret has been removed, and the greenhouse canopy has been extended aft. A rear-facing crewman controlled the target sleeve, reeling it out at a safe distance when fighter aircraft came in for gunnery practice. Under the wing is the AN/APS-4 radar pod used for searching for surface ships and submarines. (USN.)

Parked on the ramp is a squadron of Lockheed P2V-2 Neptunes belonging to VP-4, the Skinny Dragons, as indicated by their tail code, SC. These were the early submarine hunters and maritime patrol bombers. They do not have the Magnetic Anomaly Detector (MAD) stinger tail boom yet, as they still have the rear tail turret, evidenced by the shadows on the ground. Under the wings are mountings for high-velocity rockets—eight for each wing. (NAMBP.)

Rounding Diamond Head for a picturesque shot is a P2V-5 Neptune coded BE 8 of VP-6 Blue Sharks, in 1952. Painted in overall dark sea blue, this plane still has the upper gun turret, but it now has the MAD tail boom. The left wing has a fuel tip tank while the right wing has a searchlight. On the shoreline, just ahead of the nose of the aircraft, is Diamond Head Lighthouse. (USN.)

Four

Willy Victor

In the days before satellites could track the movements of potential threats against the United States by land, sea, or air, there was only the Distant Early Warning (DEW) Line, radar-picket destroyers, and the Willy Victors to warn of a possible attack. The DEW Line was a series of radar stations situated in the far northern reaches of Canada, Greenland, Iceland, and the Aleutians near the Arctic Circle. They were set up to detect incoming Soviet bombers or any land-sea invasion during the early years of the Cold War.

Supplementing the DEW Line were radar-picket destroyers spread out roughly from the Aleutian Islands to the Midway Islands. Complementing those were the long-range radar aircraft known as the Willy Victors. Based on the Lockheed Constellation series of commercial airliners, they were modified to carry massive radar systems in their belly and tall dorsal antenna housings giving them their distinctive profile. The early versions were designated WV-1 and WV-2 and were operated by VW-1 out of NAS Barbers Point.

Until satellite surveillance technology became available, these aircraft and the men who flew them had to endure long, tedious missions lasting from 12 to 16 hours, seven days a week, 365 days a year. Working with bulky equipment primitive by today's standards and extreme weather conditions from tropical heat to near freezing temperatures, men and machine were pushed to the limits.

A VW-1 (coded TE 4) PB-1W is pictured here on approach to NAS Barbers Point in October 1951. The PB-1W was the Navy's designation for the famed Boeing B-17 Flying Fortress of World War II. There were 48 B-17Gs converted to carry the APS-20E search radar under the belly. Although all the guns have been removed, TE 4 still has its chin turret under the glass nose. (VW-1 Association.)

Pictured here is another PB-1W (TE 2) landing at Barbers Point on October 15, 1952. Painted in overall dark sea blue, its previous markings have been painted out, originally bearing the modex of ND 6. In order to save weight, since they were no longer needed, all the machine guns have been removed, including the top turret behind the cockpit and the lower ball turret. By this time, the Navy was looking to replace these aircraft with one that could carry larger radar systems. (VW-1 Association.)

Pictured here is a VC-11 crew wearing flower leis (a Hawaiian tradition for visitors) upon arrival at Barbers Point in June 1952. George Stewart, who contributed many of these photographs, stands at far right. On June 18, 1952, VC-11 was disestablished, and all aircraft and personnel were transferred to VW-1. (VW-1 Association.)

A PB-1W of VW-1 sits on the ramp at Barbers Point in 1952. TE 17 still has the upper turret ring fairing above and behind the cockpit. All of these PB-1Ws started life as factory-fresh B-17Gs and never saw combat before their conversion. These aircraft lacked creature comforts such as a pressurized cabin or heating; they were designed for war, where carrying weight other than bombs was not a priority. (VW-1 Association.)

On June 18, 1952, Comdr. Fernald P. Anderson (far right), the first commanding officer, inspects the personnel of the newly commissioned VW-1 at NAS Barbers Point. At this time, Hawaii was still a US territory. It became the 50th state on August 21, 1959. (VW-1 Association.)

A PB-1W crew poses under the nose of their plane at Barbers Point sometime in 1954. Only three have been identified: Gene Bryant AD2 (first row, third from left), Pappy Herzig AD1 (first row, fourth from left), and Al Dalton AL1 (first row, fifth from left). (USN.)

Pictured here is a banquet held on July 2, 1953, celebrating the one-year anniversary of the commissioning of VW-1. At left is Comdr. W.E. Gaillard, executive officer of Barbers Point, next to Comdr. Fernald P. "Flip" Anderson. (USN.)

This photograph was taken at Crew Chief School, held from May to July 1954 in Burbank, California. It is not too hard to spot the VW-1 guys from Hawaii wearing aloha shirts among the VW-2 personnel. Pictured in the floral shirts are, from left to right, Louis Acosta, George Stewart, James Yancy, and James Bartlett, all AD1s. The rank of AD in the Navy stood for aviation machinist mate, and the "1" indicated first class. (VW-1 Association.)

By the early 1950s, the Navy was looking for another aircraft to replace the PB-1W. As the missions expanded, the B-17 airframe was taxed to its limits. The Navy chose the Lockheed Constellation. Originally designed as an airliner, it had the long range and ability to carry two large radar systems and related equipment. Here, Lockheed's L-1049 Super Constellation prototype makes an early test flight over California around 1951. (VW-1 Association.)

Albert Dalton AL1 (left) and Bill Lukeshek AL3 wear cold-weather exposure suits during their training prior to deployment to Japan and Korea in January 1953. During flight, these suits were somewhat uncomfortable but necessary for surviving the near-freezing temperatures in the waters off Korea. (VW-1 Association.)

Crews are testing the exposure suit in the swimming pool at NAS Barbers Point in January 1953. Water-survival training was conducted in the pool for all climate conditions. Aircrews had to be certified with all survival equipment and flotation devices prior to deployment overseas. Although the tests are conducted in ideal conditions, they were necessary should ditching the aircraft or bailing out over the ocean become necessary. (Both, VW-1 Association.)

Pictured here in December 1952 at Barbers Point is the second PO-1 built; it would be redesignated WV-1. This airplane (TE 1) had the smaller belly radome. (Courtesy of the Anderson family.)

Dated April 16, 1954, this is the third production WV-2 in flight off the Waianae Coast near Barbers Point. The massive radomes above and below the fuselage housed the AN/APS-45 height-finding radar (top) and the AN/APS-20 air search radar in the ventral dome. The aircraft is bare metal, with only the center section painted light gray. (USN.)

This VW-1 flight line in the fall of 1954 shows the difference in size between the PB-1W and WV-2. The Constellation, or Connie as it was known among Navy personnel, could carry the massive radars and had a tricycle landing gear arrangement, while the smaller PB-1W/B-17 was a tail dragger. (USN.)

Pictured here in 1955 is a VW-1 Willy Victor coded TE 1 (BuNo 131388) on the ramp at Barbers Point. VW-1 was tasked with proving the viability of the Willy Victor in the early days of Airborne Early Warning. Their work paved the way for modern radar systems. (VW-1 Association.)

A VW-12 Constellation coded SH 25 flies past picturesque Diamond Head Crater in the mid-to-late 1950s. VW-12 was a part of several squadrons forming Airborne Early Warning Barrier Squadron Pacific (AEWBARRONPAC). (VW-1 Association.)

This is the radar operator's station inside a Willy Victor. Before transistors or microchips, large vacuum tubes did much of the electronic information processing, primitive by today's standards but state-of-the-art technology in the 1950s. (VW-1 Association.)

Looking aft in a WV-2 interior, this is the CIC (combat information center) section, the heart of the radar plane. To operate the radar equipment there can be anywhere from 11 to 25 crewmen depending on the mission. Typically, there was a flight crew of six as well. (VW-1 Association.)

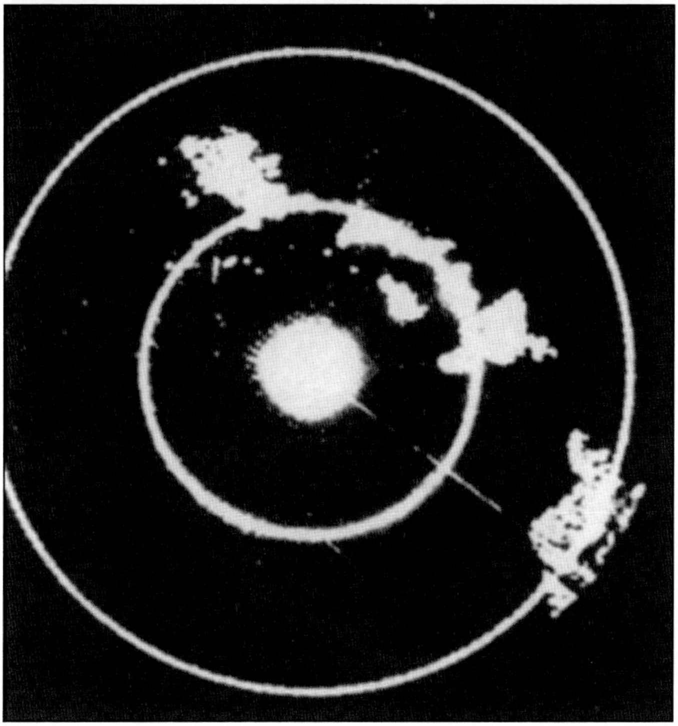

This radar image from an altitude of 10,000 feet shows the islands of Hawaii; from upper left are Oahu, Molokai, Lanai, Maui, and the Big Island. (Willy Victor Group.)

Pictured here are a pilot and co-pilot in a Willy Victor cockpit. For its size, the Constellation cockpit was fairly small; note the overhead control panel and round dial instruments. (VW-1 Association.)

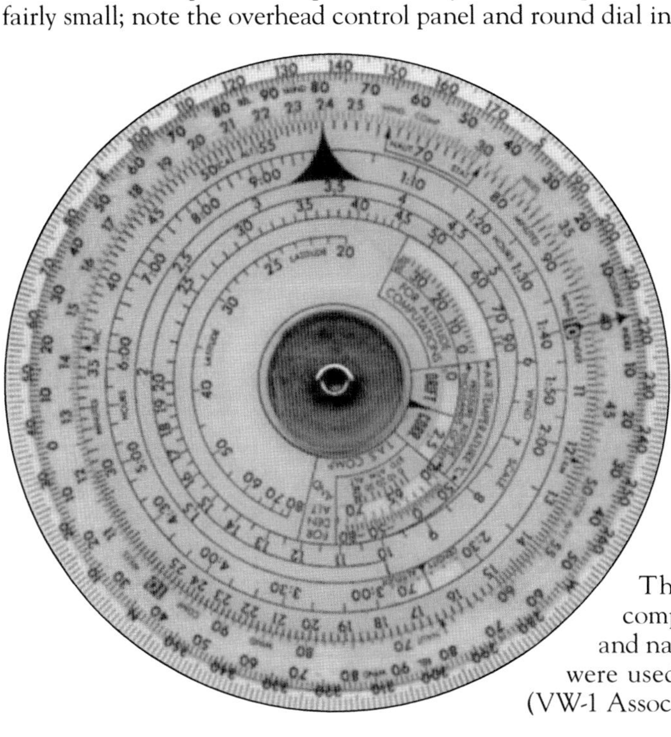

This is a 1950s–1960s "laptop computer" for pilots, flight engineers, and navigators. In those days, slide rules were used to design and build airplanes. (VW-1 Association.)

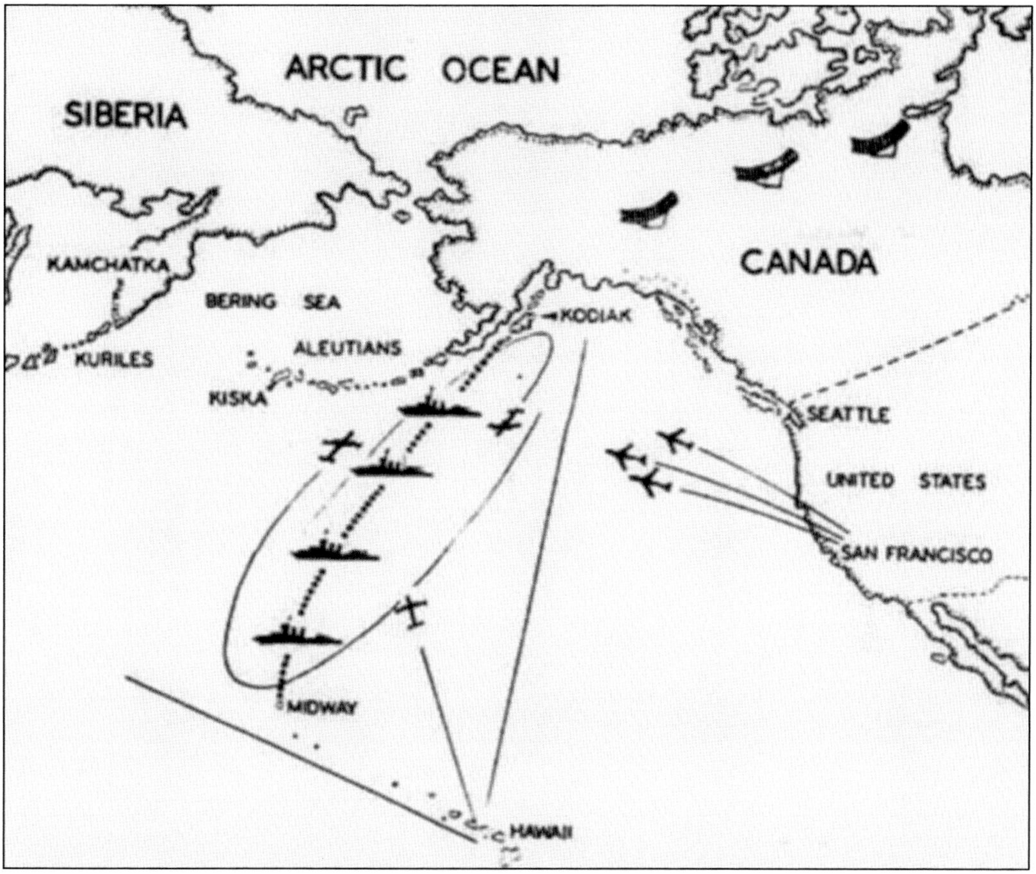

The radar stations in northern Canada were near the Arctic Circle. Forming a defensive line from the Aleutians to Midway were the radar-picket destroyers. Willie Victors flew a racetrack pattern around the destroyers. Four to five aircraft were required to cover the line and supplement the destroyers. (Willy Victor Group.)

Shown here is the AEWBARRONPAC squadron at Barbers Point around 1960. These are WV-2s (original designation PO-2W) with new paint schemes. The forward operating bases were Midway Island and Agana, Guam. These aircraft were constantly flying to and from their patrol area, providing coverage 24 hours a day, seven days a week, 365 days a year. (USN.)

Pictured is the AEWBARRONPAC hangar and flight line in 1962. Three WV-2s and two R7V-2s (without radomes) are being serviced. The R7V-2 was the turbo-prop version of the Connie and was used for cargo and troop transport; only four of these were built. (USN.)

This EC-121K was photographed from the tower at Barbers Point in the early 1960s. TE 5 was built in March 1958 and served 12 years until it was struck from the US Navy inventory on July 28, 1970. After being placed in storage for several years, it was scrapped on May 24, 1976. (USN.)

By 1962, US Navy aircraft designations were changing; WV-2s were now EC-121Ks. BuNo 143223 (SH 3) is parked next to the tower at NAS Barbers Point on March 4, 1965. By September 1965, barrier force operations ceased. However, the Connies continued to serve throughout the Vietnam War as electronic surveillance aircraft. (USN.)

Pictured here is an original VW-1 patch from 1952–1953. The globe or circle was light blue with Paul Revere holding a yellow glowing lantern and riding a white winged horse. (VW-1 Association.)

Although the barrier force missions had ceased in 1965, the EC-121s continued to serve in other capacities. Although the missions were long, tedious, and less glamorous, it was not without peril or danger. On April 15, 1969, a EC-121M on a reconnaissance mission was intercepted and shot down by North Korean MiG-17s over the Sea of Japan. All 31 Americans onboard the unarmed aircraft were killed. (VW-1 Association.)

Five

Neptune Rising

With the end of World War II, and the Cold War escalating, naval warfare was evolving. There would no longer be ship-to-ship conflict on a massive scale. Instead, the threat would come from below the surface of the sea in the form of submarines. As the Atomic Age dawned, the two seemed destined to go together—delivery of an atomic weapon via submarine.

Developing the technology seemed daunting but not impossible. In the early 1950s, it was still a new concept. The United States and the Soviet Union embarked on a crusade to build massive fleets over the years, with the emphasis on submarines.

Long-range maritime-patrol aircraft, so valuable in World War II, would be needed in the foreseeable future. When the Lockheed P2V Neptune arrived, it was a suitable platform for the role of maritime patrol. New antisubmarine warfare (ASW) detection equipment was continuously developed and installed on the new patrol bomber. The first production variants still looked like classic World War II bombers with gun turrets. These gave way to the new submarine detection gear such as the magnetic anomaly detector, or MAD.

Within 10 years of service, the P2V Neptune started showing its limitations; as more equipment was placed onboard, the aircraft required jet engine pods for takeoff, as its weight became a liability.

A VP-22 (coded QA 1) P2V-5F (BuNo 131429) flies past Diamond Head in late 1958. By this time, the P2V Neptune had gone through numerous upgrades, with more to come. (USN.)

QA 1 has climbed to altitude and executes a wing-over maneuver for the camera ship. The large belly dome housed the AN/APS-20 search radar. Due to the weight of the detection gear, the gun turrets were moved to compensate for the additional weight. Even with that tradeoff, it still needed jet pods and sometimes RATO bottles (rocket assisted takeoff) on the rear fuselage sides. (USN.)

A P2V maritime patrol bomber sits on the ramp at NAS Barbers Point near a squadron of F9F-2 Panthers from the USS *Essex*. They were on their way to Korea in this photograph from around 1951. (Townsend.)

Pictured here is a front view of the same Neptune, a P2V-3 evidenced by its three-bladed propeller. The small bumps under the wings (eight on each side) are mountings for unguided rockets. (Townsend.)

A Douglas R5D military transport of VR-21 is being serviced and fueled. This squadron provided fleet support by air overseas. (Townsend.)

This is a rear view of another VR-21 R5D near the operations building and control tower. The cranes were used to remove aircraft that made a "wheels up" or belly landing on the runway. (Townsend.)

A number of these photographs were donated by the family of Eugene "Gene" Townsend, who served with VR-21 at NAS Barbers Point between 1951 and 1953. Pictured here are, from left to right, Gene Townsend, Rudy Altenburg, Bill Trengove, and Johnny Longwill (kneeling). (Townsend.)

The Base Exchange Department, or the general store, was a favorite place to go on base and the location of both photographs shown on this page. (Townsend.)

Shipmates of Gene Townsend enjoy a day at the beach, complete with seaweed grass skirts. Pictured here from left to right are Rudy Altenburg, Bill Trengove, and Jack ?. (Townsend.)

This is a formation of VP-4 P2V-5F Neptunes near Oahu in 1957. The overall dark sea blue paint scheme used since World War II was changed to a two-tone dark gray and white. (NAMBP.)

This is an oblique aerial view of the air station with the now-shuttered Ewa Field nearby. The runways at Ewa Field were deemed too short for jet aircraft, which required a longer takeoff and landing roll. The dark-colored road to the left is the railroad track. (NARA.)

Pictured here in 1958 is the main gate to the air station. Above the guard shack is "Merry Christmas and Happy New Year" in Hawaiian. (J.G. Klein.)

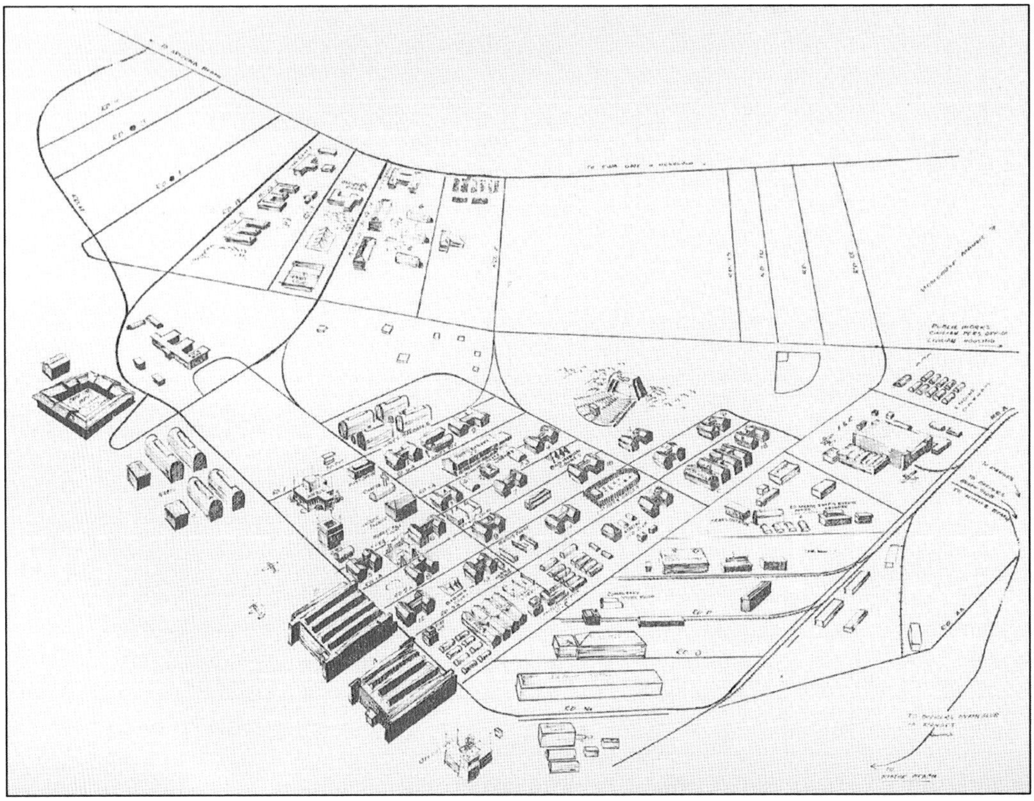

This hand-sketched directional map showing the layout of NAS Barbers Point was printed and provided to visitors. In the lower center are Hangars 110 and 111. (Cislo.)

Shown here is the Douglas R4D-8 (later redesignated C-117D) cargo plane. This is the Navy version of the famed C-47 troop carrier of World War II. This plane is doing an engine run-up test, as much of the nacelle panels have been removed. (USN.)

A Douglas R6D has its propeller primed before engine start by two sailors for the plane's first maintenance check on November 11, 1952. (USN.)

Shown here is the squadron flight line of VP-6 Blue Sharks in late 1959 or early 1960. The Neptunes have received the new dark gray and white paint scheme. The white roof helped keep the cabin cool by reflecting the bright Hawaiian sun. (USN.)

Pictured is a VP-4 Skinny Dragons SP-2H, an improved version of the P2V-7 with upgraded ASW/ECM gear, flying over the Pacific on October 15, 1963, while on deployment to NAF Naha, Okinawa. (USN.)

Another VP-4 Neptune rounds Diamond Head as it returns to its home base, NAS Barbers Point, around 1963. The tropical weather has taken its toll on the paint finish, as it is starting to show some oxidation and wear. Salt, air, and sun kept the airframe-maintenance crews busy in dealing with corrosion. (USN.)

QA10 and 7 from VP-22 fly in formation over the Ewa Plain in 1960. The right wingtip houses the powerful AYQ-2A searchlight operated by a crewman in the nose. (NAMBP.)

In an earlier photograph of VP-22s, the flight line is wearing the overall dark sea blue paint, as well as their old tail code CE. (NAMBP.)

Pictured here in February 1966 is a VP-1 SP-2H in the new light gull gray and gloss white scheme. As Navy aircraft became capable of delivering atomic or nuclear weapons, it was thought that the dark paint schemes would absorb the intense flash and ultraviolet light of a nuclear explosion. To some degree, the lighter colors would reflect the light. (Tailhook Association.)

Pictured here is YB 5 of VP-1 Screaming Eagles in flight during a Market Time mission off the coast of Vietnam on April 13, 1966. During the Vietnam War, Operation Market Time was set up to intercept North Vietnamese gun-running trawlers and small boats. (USN/R.D. Moser.)

Six

Home Front

Since its beginning, naval aviation has always evoked a romantic, if not glamorous, allure. No doubt Hollywood perpetuates that vision of pilots in flight suits climbing in fighters or bombers, taking off on daring missions, or standing in crisp dress uniforms courting the ladies.

The truth is, beyond the mystique and glamour is a straightforward military service steeped in ritual and tradition. After the aircraft are parked, equipment stowed, and flight suits hung, the daily drudgery of maintenance, paperwork, and much more mundane, but vital, work begins. At the end of the day, family or personal life begins; time with loved ones or friends is always a welcome pleasure. For those on deployment in some far-off land, those times are few and far between.

For those dependents growing up on a military base, the time spent there can be enjoyable or a sheer nightmare. The military does its best to provide for the families and those who serve in uniform. Housing, recreation, cost of living allowances, and benefits are some of the tangible and intangible things that make living in a place like NAS Barbers Point enjoyable, or at least bearable.

Sitting in his 1960 Triumph Herald is Lt. Comdr. Alex Sundstrom and his children, around 1962. The car is painted in Litchfield green, and their home was located off-base in an area known as Iroquois Point, near the entrance to Pearl Harbor. (Sundstrom.)

Lieutenant Commander Sundstrom in dress whites stands with his son David on Armed Forces Day, May 19, 1962. Behind them is the Navy version of the Lockheed T-33 Shooting Star, designated the TV-2D. (Sundstrom.)

Pictured here on the same day is a Douglas AD-5 Skyraider. VU-1 aircraft did many tasks, including towing target sleeves or banners for gunnery practice. The wire reel used for towing the targets is mounted on the belly of the aircraft. (Sundstrom.)

Pictured here is an ejection-seat demonstration during an air show around 1965–1966 at NAS Barbers Point. The aircraft is a Grumman F9F-5KD Panther; in spite of the damage the seat might cause, the aircraft was repaired and returned to flying service. (Nankivil.)

Sporting a shark mouth, a VU-1 Vought F-8A Crusader cruises past Diamond Head in 1966. The vertical fin was painted red with the tail codes in black with a white outline. (USN.)

Standing by the landing gear of an AD-5 Skyraider are Lieutenant Commander Sundstrom's children Bob (left), David, and Betty. (Sundstrom.)

Betty and David sit in the cockpit of a Lockheed TV-2D. Many people remember their first time sitting in a military jet as exciting but a bit scary for young children. (Sundstrom.)

Bob Sundstrom sits on a Radioplane KD2R5 Shelduck, a basic training target drone. The stencil on the nose reads, "Cash Reward Authorized for Return of Target Aircraft. Phone VU-1 4711 EXT. 64202 or 64144." (Sundstrom.)

In the original color photograph, 11-year-old Betty Sundstrom wears a matching red dress next to this Regulus missile. Built by Chance Vought, the Regulus was an early nuclear-armed cruise missile. (Sundstrom.)

AD-5 Skyraider UA-53 starts its engine for a maintenance check. The heavy exhaust stains along the side of the fuselage were typical of all Skyraider variants. (USN.)

On February 8, 1963, at NAS Barbers Point, Comdr. George Smith gives an inspection of VU-1. Lt. Comdr. Alex Sundstrom (center) is flanked by Lt. John Muckenthaler (right) and Lt.(jg) Clint Johnson. Two years later, over the hostile skies of Vietnam, Clint Johnson would be in the midst of the only prop-versus-jet aerial engagement of the war. (Sundstrom.)

On June 20, 1965, Lt. Clinton B. Johnson along with Lt. (jg) Charles Hartman (wingman), Lt. (jg) Jim Lynne, and Lt. Comdr. Ed Greathouse (flight leader) were supporting a search-and-rescue mission for a downed USAF F-105 pilot when their flight of four A-1H Skyraiders encountered two North Vietnamese MiG-17 jet fighters. In the ensuing dogfight at nearly treetop level, Johnson and Hartman fired their 20-millimeter guns as one of the MiG-17s flew directly at them. Both Johnson and Hartman were credited with the kill. This is Lieutenant Johnson's Skyraider from VA-25 (NE 577) on board the USS *Coral Sea*. (USS Midway Association.)

Here the pilots recount the engagement for their shipmates in the ready room. Pictured from left to right are Charles Hartman and Ed Greathouse giving "pilot talk" while Clint Johnson (arms folded), Jim Lynne, and others listen in. (USS Midway Association.)

Shown here is a MiG-17 fighter jet in a typical camouflage scheme used in Vietnam. As Alex Sundstrom relates, "VU-1 was the perfect incubator for combat pilots." He further states that the unit tracked torpedoes, operated aerial drones, and studied target acquisition through aerial photography and bomb damage assessment—just some of the principal elements for developing situational awareness in a high-stress combat environment. (US Air Force Museum.)

At NAS Barbers Point during the late 1950s and early 1960s, a USAF B-52 would fly from the air station carrying a live nuclear weapon to be dropped on Johnson Atoll, and Kiritimati (Christmas Island) Kiribati. It was called Operation Dominic. Many residents old enough can still recall the strange, but spectacular, sunsets the nuclear explosions created. This is the crew and the B-52A Stratofortress at Barbers Point. (NAMBP.)

For many, especially for the officers and their families, living on the base was idyllic, as these plantation-style homes suggest. The grounds were very well kept, with large shade trees planted around the property. (NARA.)

This is a view of another officer's house. These homes were built in the late 1940s and continually remodeled over the years to keep up with the changing times. (NARA.)

This aerial view of the air station was captured on March 9, 1964. The officers' homes in the previous photographs were located in the lower-right corner. Only a few are left and occupied; the rest have been left abandoned and burned down by arsonists. Enlisted quarters are located toward the center in most of the long buildings. (USN.)

A typical sight on any military installation are the enlisted men's and women's barracks. Pictured here in the mid-to-late 1960s is one of the main enlisted quarters located not too far from the flight line. (Bremus.)

One of the many sailors on board NAS Barbers Point was Bob Bremus, looking squared away in dress whites around 1967. (Bremus.)

In the 1940s, the Olympic-size swimming pool was built for base personnel and their dependents and was also used to train aircrews for water survival. (Bremus.)

Right across from the recreation center and swimming pool was the base chapel. (Bremus.)

This view of the control tower around 1966 was taken from under the wing of a C-118 Liftmaster of VR-21. (USN.)

Pictured is a VR-21 Fleet Logistical Support Douglas C-118B on the ramp around 1965. These planes were used for personnel transport, and BuNo128424 was converted to a VIP transport with more luxury amenities fitting for high-ranking officers. (USN.)

Pictured here is the front of the air terminal. On either side of the "Aloha" sign at the entrance were two long surfboards. (NAMBP.)

C-118B coded RZ 604 has its cargo loaded by one of the air terminal crew. According to Bob Bremus, the air terminal was constantly busy during the Vietnam War, as Hawaii was the crossroads, and there was a constant flow of men and material. (Bremus.)

Two of Bremus's coworkers and shipmates were Ken Brown (left) and a Native American named Romero they called "Tonto," which he did not mind. The cargo building/air terminal is to the right, and the building beyond is the photography lab, where all reconnaissance photographs were processed. (Bremus.)

Bob Bremus handled all the clerical work for the chief petty officer in charge of the air terminal. According to Bremus, it was a fun time for him in Navy at Barbers Point with lots of fond memories on and off the base. Things would change for him in 1969. (Bremus.)

On January 14, 1969, tragedy struck, as an explosion and fire occurred aboard the nuclear-powered aircraft carrier USS *Enterprise* (CVN-65). As her aircraft were getting ready for launch, a Zuni rocket pod overheated and cooked off due to an aircraft starter unit's hot exhaust. As the rocket shot off, it struck another aircraft, starting a chain reaction of fires and other rockets and bombs going off. The fire destroyed 15 aircraft, 27 sailors lost their lives, and another 314 were injured. (USN.)

With gaping holes on the flight deck and the remains of at least a dozen aircraft and debris strewn about, the USS *Enterprise* steams toward Pearl Harbor. After repairs, the mighty carrier returned to home port with Bob Bremus on board as one of many replacements for the 341 total casualties sustained. (USN.)

Seven

THE FABULOUS 1960S

The young, new president showed much promise; it was the beginning of a new decade full of hope and dreams unimaginable in the decade before. The proclamation was made for putting a man on the moon before the decade was out. The goal was very bold, but held much excitement at the prospect of achieving the impossible. As Pres. John F. Kennedy stated, "We choose to do these things not because they are easy, but because they are hard."

The 1960s seemed like another golden era in America's history; things were going to be different. However, the Cold War loomed and promised to reach a boiling point somewhere in the world. The Cuban Missile Crisis, rising tensions in the Middle East, and Communism spreading through and engulfing third-world countries were all pushing toward a nuclear confrontation between the superpowers.

One of those hotspots that directly involved America was a little-known country then—Vietnam. It was a world away for many Americans, civilian and military alike. At NAS Barbers Point, life went on; it still had to continue its job, even accommodating something never before seen—Hollywood.

Pictured here in front of the control tower, a Boeing N2S-5 Stearman biplane and a North American FJ-4 Fury exemplify the tremendous leap in naval aviation. Separated by merely two decades, the technological advances in aeronautical design and construction would have been mindboggling to the original designers of the biplane. (World War II Database.)

A line of FJ-4 Furies of VU-1 sit chocked and tied down. VU-1 provided training for pilots in the fleet to keep up their proficiency in combat readiness. The word utility suggests something unglamorous, but what they did was essential for fleet pilots. (USN.)

A Vought SSM-N-8 Regulus missile is prepared for launch by the ground crew. The Regulus could be launched by submarine or surface ship and was deployed from 1955 to 1964. To help get it into the air, two large booster rockets were fitted on the aft end of the fuselage. (USN.)

With the booster lit, the Regulus takes flight; in a few seconds, its main engine will ignite. The Regulus launches took place on the island of Kauai at the Pacific Missile Range Facility (PMRF), also known as Barking Sands. (USN.)

As the Regulus Missile gained altitude, F-8 Crusaders from VU-1 would intercept and escort it until its fuel ran out or guide it into the ocean. (Nankivil.)

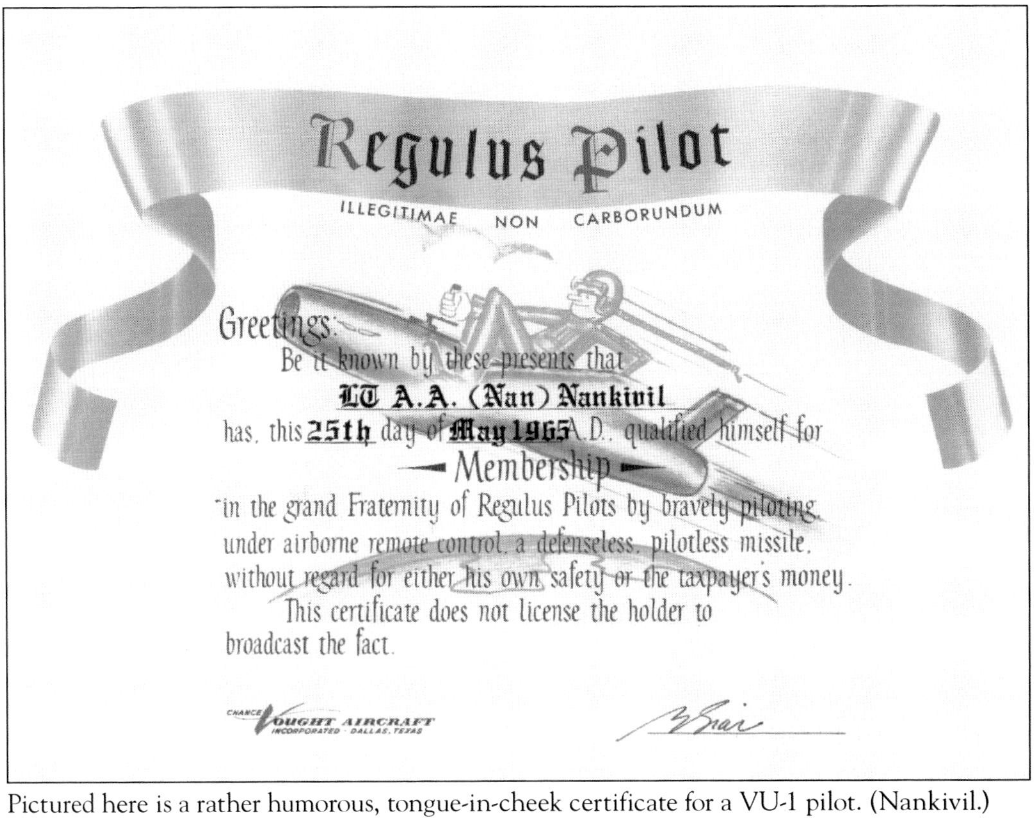

Regulus Pilot

ILLEGITIMAE NON CARBORUNDUM

Greetings:

Be it known by these presents that

LT A.A. (Nan) Nankivil

has, this 25th day of May 1965 A.D., qualified himself for

— Membership —

in the grand Fraternity of Regulus Pilots by bravely piloting, under airborne remote control, a defenseless, pilotless missile, without regard for either his own safety or the taxpayer's money.

This certificate does not license the holder to broadcast the fact.

CHANCE VOUGHT AIRCRAFT INCORPORATED · DALLAS, TEXAS

Pictured here is a rather humorous, tongue-in-cheek certificate for a VU-1 pilot. (Nankivil.)

This publicity shot shows the nose of a VU-1 Crusader in front of the control tower in the mid-1960s. The tower itself has been modernized with new glass panels and radio antennas. (USN.)

On July 1, 1965, Utility Squadron One (VU-1) was redesignated Composite Squadron One (VC-1, and changed its name from the Unique Antiquers to the Blue Aliis (Warriors). This photograph dated July 26, 1970, shows a Vought F-8K Crusader. The Hawaiian Warrior's head design was inspired by a local brewery known as Primo Beer; VC-1 obtained permission to use the logo, as it was almost free advertising, so the story goes. (USN.)

In 1969, VC-1 traded in its F-8 Crusaders for the Douglas A-4 Skyhawk. Also dated July 26, 1970, this photograph shows two A-4C models. In time, other versions of the Skyhawk appeared, including the earlier B model, the long-nose E model, and the two-seat trainer versions, the TA-4J and TA-4F. (USN.)

As a composite squadron, VC-1 operated other types of aircraft as well, including rotary-winged, or helicopters. Photographed on the same day as the previous images is a Sikorsky UH-34D Seahorse finished in overall seaplane gray. The object on the side of the fuselage is a high-frequency (HF) antenna. (USN.)

UA-23 and UA-24 cruise in a formation during a training flight in 1969. The H-34 Seahorse served in the early stages of the Vietnam War, mostly with Marine Corps units. (USN.)

After the order came out to repaint all Navy aircraft in the new light gull gray and white, VC-1 and all other units did so. UA-23 appears in the new paint scheme in early 1971. The fuselage band was blue (FS 15090), with stars now added around the warrior's head. The blue field represented the Pacific Ocean, and the yellow stars signified the eight Hawaiian Islands. (USN.)

This squadron photograph of VC-1 was taken on June 5, 1965, for a change-of-command ceremony. They are pictured with the three principal fixed-wing aircraft they operated. (USN.)

Comdr. Richard T. Manning (center), commanding officer of VC-1, awards aircrew wings to some of the enlisted personnel. Clockwise from top left are Porter, Manning, Litchfield, Lile, and Atwell. (Atwell.)

"Just a little higher, now hold it right there!" VC-1 target drone crewmen are getting No. 48 ready under the wing of a Lockheed DP2V Neptune. The target drone is a Ryan Firebee BQM-34A. (Atwell.)

The Neptune could carry two drones under its wings, and even though they were for target practice, every effort was made to minimize destruction so they could be recovered and used again. (Atwell.)

The Firebees were painted bright orange to make them easier to spot. Later models were fitted with various control systems to emulate fighter-like maneuvers, and radar emitting devices to simulate the radar signatures of enemy aircraft. (Atwell.)

This is a forward port quarter view of UA 30, this time sporting the VC-1 blue Aliis insignia on the nose. When the drone was hit, a parachute would open, and if possible, it would be snatched from the air by helicopter. The Firebee could float for a time if it ditched in the ocean. (USN.)

Comdr. W.C. Dukes, commanding officer of VC-1, is pictured here conducting a squadron inspection on December 11, 1968. On the right, being eyed by Commander Dukes is W. Russell Atwell, who contributed many photographs for this chapter. And yes, they passed inspection. (Atwell.)

No, this is not a vintage World War II photograph. Hollywood came to NAS Barbers Point for the filming of the movie *Tora! Tora! Tora!*, the story of the attack on Pearl Harbor on December 7, 1941. Filming of the aerial sequences started in late summer 1968 through the spring of 1969, about eight months. The film was released in 1970. (Cislo.)

Pictured here is one of five B-17s used for the film, which all came from a company in Arizona and had to be modified to look like World War II bombers in 1941. All of the American and Japanese planes were parked at NAS Barbers Point. The B-17s portrayed Maj. Truman Landon's flight from Hamilton Field, California. In real life, like the movie, the Japanese attack force was mistaken for Major Landon's group. (Atwell.)

This is actually a North American AT-6 Texan convincingly modified to look like the Aichi D3A-1 Val dive bomber. Note the dummy rear gunner in the cockpit. Civilian and active duty military pilots were hired to fly the aircraft. (Atwell.)

This image is looking east at the entire Val dive-bombing force with Hangar 111 in the background. In the next row behind them are the Kate high-level and torpedo bombers. (Atwell.)

These pseudo-Japanese planes are in reality the Vultee BT-13 World War II trainer heavily modified with parts from the AT-6 Texan to look like the Nakajima B5N2 Kate torpedo bomber. Note the prop torpedo on the ground. (Atwell.)

Comdr. Louis Cislo, an active duty naval aviator and World War II veteran, flew the lead Japanese bomber in the film. His aircraft sported a red- and yellow-striped tailfin, just like the plane of the actual strike leader, Comdr. Mitsuo Fuchida. (Cislo.)

Lt. Louis Cislo stands by his Grumman F6F-5 Hellcat on Saipan in 1945. It is ironic that less than 25 years later, Lieutenant Cislo would portray the very enemy he was trained to shoot down. (Cislo.)

Eight

Orion the Hunter

Russia was building a sizable fleet of submarines that would potentially pose a threat to America and her naval forces. Every means was being developed to counter the submarine threat: surface ships, other submarines, aircraft carrier–borne crafts, and land based aircraft, too. One of the few consolations was that the Soviet Union had only diesel-powered subs in the early 1960s, which meant they had to surface to recharge their batteries and expel poisonous and explosive gasses.

The United States was the first to invent the nuclear-powered submarine, which sent a shockwave through the Kremlin. The USS *Nautilus* had a tremendous advantage over diesel powered submarines in that it could stay submerged almost indefinitely. That, coupled with the ability to deliver atomic and then later nuclear weapons, sent shivers through all nations. Such a submarine could launch ballistic missiles without being detected.

A new submarine-hunting platform was needed to go out far enough before an enemy submarine could unleash its nuclear payload within range. The P2V Neptune was showing its age. The lack of range and its size limited the amount of detection it could carry for the growing mission requirements. Something was needed quickly, without spending millions of dollars and years of developing a new aircraft. Once again, Lockheed provided the solution in the form of a nearly failed airliner known as the Electra. The Navy saw great potential and transformed it into the P-3 Orion, named after the celestial hunter.

On September 7, 1965, VP-22 passes Diamond Head, marking its arrival in Hawaii as the first operational P-3 Orion squadron in the islands. After completing transitional training to the new type, VP-22 had flown the PV2 Neptune since 1950. This photograph was taken from a P2V Neptune coded QA 12, whose wingtip fuel tank can be seen at left. (USN.)

Pictured here is the changing of the guard at NAS Barbers Point. With props squared, QA 1 sits next to its predecessor, the P2V-7 (QA 6). The Navy now had the range and better capability, but it would take years for the airplane to fully mature as the preeminent maritime patrol bomber. (NAMBP.)

Deliveries of the new sub hunter began to accelerate following its introduction in June 1962. Since the production line for the airliner version was already in place, it was fairly easy to produce the new sub-hunting version, thereby avoiding large-scale tooling costs. In this photograph, a P3V-1 from VP-28 shadows and is about to pass a US diesel-powered submarine returning to Pearl Harbor. (USN.)

Looking north in this photograph taken on October 4, 1965, is the air station as it looked at that time. It has grown considerably over the years since World War II, and the runways have been repaved several times to accommodate heavier aircraft. Among other improvements was updated striping on the runways and taxiways. (USN.)

Photographed over the Sea of Japan during its 1986–1987 deployment is a VP-1 P-3C (YB 1) on a low-altitude mission. After detecting a possible submarine contact with its MAD gear, the P-3 would drop a series of sound-emitting sonobuoys to pinpoint the exact location. Often, the P-3 would go to sea level and visually look for a submarine riding just below the surface or at periscope depth if ocean and weather conditions were favorable. (Mahlon K. Miller.)

The rapid development of Russian submarines prompted the US Navy to find ways to counter the threat. The Soviet navy had the capability to deliver nuclear weapons via ballistic missiles close enough to the United States where reaction time was minimal. This Golf II–class submarine had three missile silos in the massive sail. The United States tried to recover a similar, but older, version of this type of submarine in the 1970s. It was known as Project Azorian, and the *Hughes Glomar Explorer* was the ship that attempted (with some success) to retrieve the Soviet submarine from the ocean floor. (USN.)

Early P-3As and B models had the AN/ASR-3 Trail Detector sniffer, used to detect the exhaust emissions of diesel-powered submarines. Even after the boat had disappeared under the waves, the gasses could still be picked up. When nuclear-powered submarines appeared, they did not have to surface like this Sierra-class attack submarine. While it did not have ballistic missile–launch capability, it posed an even greater threat to the US Navy's carrier battle groups. (USN.)

Should the P-3 be cleared to fire upon a submarine threat, one of the weapons at their disposal is the Mk-46 acoustic homing torpedo. With a speed of 40 knots (45 miles per hour) its 500-pound warhead could punch a hole in the titanium hull of Soviet submarines. Death would be instantaneous, as water would rush in at blinding speed. (VP-4 Association.)

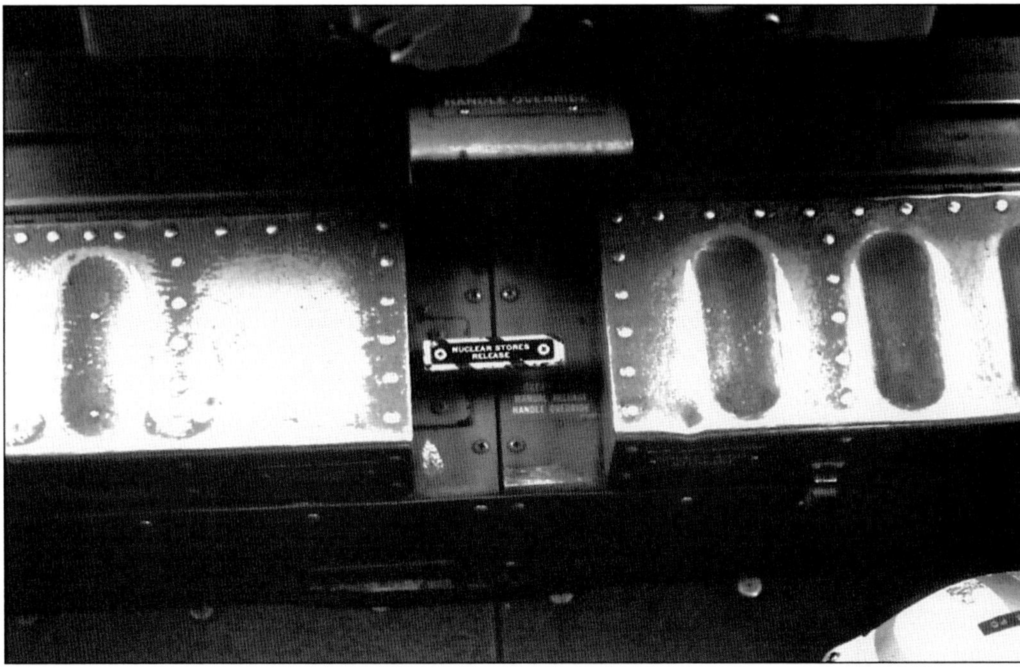

The nuclear weapons release handle was located on the cockpit floor in front of where the flight engineer sat, between and just behind the pilot and co-pilot. If it came to that point, they could and would have—thankfully it did not. (Ruisch.)

A VP-4 P-3B (bureau number 153433) is photographed in flight on July 19, 1978. Beneath the wing is a high-powered searchlight. The jet-style engine design over the wings gave the P-3 seven percent additional thrust, which allowed the aircraft to shut down one or two engines to conserve fuel for long patrol missions.

Nine

The Super 1970s and 1980s

Being the only Navy air station in the Pacific, NAS Barbers Point was an important base for the maritime patrol bomber fleet. The Cold War showed no signs of ending; there were numerous times when the superpowers came to the brink of all-out nuclear war. There was growing concern among the top military leaders over whether other nations in the Pacific region who were friendly with the United States could be called upon to actually work together in a maritime capacity. To answer that question, a large-scale multinational maritime training exercise was developed.

RIMPAC, or Rim of the Pacific, was the answer. A biannual maritime exercise tested the strengths and weaknesses of several nations working together on all levels, including land, sea, and air scenarios. Beginning in 1971, that inaugural exercise saw Australia, New Zealand, Canada, and the United Kingdom working together for the first time since World War II. Since then, RIMPAC has grown to include a host of other nations, including Japan, South Korea, Mexico, South America, and others in the Pacific region. NAS Barbers Point played an important role in hosting the patrol squadrons of these nations.

Here is an aerial view of the new hangar at the Coast Guard Air Station, Barbers Point, in 1970. The Coast Guard search-and-rescue aircraft were continuously keeping up with the times, from the early Grumman HU-16 Albatross to the Lockheed C-130s. (USCG.)

This photograph of the Coast Guard air station in the mid-to-late 1960s shows the older wooden structures (long buildings), which were torn down to make way for the new hangar in the preceding photograph. In the far distance is the oil refinery at Campbell Industrial Park. Between that and the end of the runway was the arming area for the B-52 carrying a nuclear bomb for Operation Dominic. (USCG.)

At some point, NAS Barbers Point acquired a second name: John Rodgers Field, or JRF, as it is widely referred to in the general aviation community. The name comes from a Navy commander and early aviation pioneer who flew the first transpacific flight with his five-man crew—an intrepid journey that nearly ended in disaster. (NARA.)

On August 31, 1925, Commander Rodgers took off from San Francisco in a flying boat, PN-9 No. 1. About 365 miles from Oahu they ran out of fuel and landed in the ocean. Rodgers and his crew took the fabric from the wings and converted the flying boat into a sailing ship of sorts. After searching for eight days, the Navy gave up. They were spotted near the island of Kauai by a submarine and rescued. Draped in flower leis, Rodgers and his crew were instant celebrities and hailed as heroes. (NARA.)

The PN-9 No. 1 flying boat was built by Naval Aircraft Factory and its design was based on the British Felixstowe of World War I. After being towed to the harbor, the plane was repaired and flown to Oahu. (NARA.)

On Oahu, Commander Rodgers and his crew received honors from Gov. Wallace R. Farrington. Two years later, an airport was named in his honor. John Rodgers Airport was later renamed Honolulu International Airport in 1947. As for NAS Barbers Point being referred to as John Rodgers Field, no official date was given, although sporadic use began in the 1970s or later. (Hawaii State Archives.)

This aerial view taken from a VP-4 P-3 Orion shows the Rainbow Hangar; all P-3s at that time had a small rainbow painted on them, usually on a fin fillet. This is the nose hangar, where all major maintenance work was done on the Orions. (Ruisch.)

ZE 1 of VP-17 White Lightnings flies near the air station in August 1971. Like many patrol squadrons, VP-17 can traces its beginnings to World War II. After having served for more than 50 years, they were disestablished on March 31, 1995. (USN.)

During the 1970s, these were the P-3 units stationed at NAS Barbers Point, the Fleet Support Squadron: L-R VR-21 C-130F; VP-1 Screaming Eagles (YB 5); VP-17 White Lightnings (ZE 2); VP-6 Blue Sharks (PC 3); VP-4 Skinny Dragons (YD 3); and VP-22 Blue Geese (QA 4). Other squadrons also stationed there but not pictured are VPU-2 Wizards (P-3 Orion); HSL-37 Easyriders (Kaman SH-2F); and Composite Squadron One (VC-1) Blue Aliis. (USN.)

This is a view of P-3s from a P-3 of VP-4 sometime in the late 1970s. During that time, nearly 80 P-3 Orions were based at Barbers Point. In the far distance is the burgeoning town of Makakilo. Between there and the air station is now the city of Kapolei. (Ruisch.)

Ten

THE 1990S AND BRAC

The breakup of the Soviet Union was unexpected and totally unforeseen. To the intelligence observers in the West, it was surreal that Communism was falling apart before their eyes; it was the end of the Cold War, or so it seemed in Europe. There were still tensions in the Middle East, with a despot invading a little oil-rich country called Kuwait. Once again, America found itself embroiled in another shooting war. When the 911 call went out to the free world, VP-1 P-3s from NAS Barbers Point were the very first responders, flying surveillance missions and anti-shipping patrols in the North Arabian Sea and the Persian Gulf.

Operation Desert Shield/Desert Storm made quick work of Saddam Hussein's military with the help of coalition forces. With the end of that conflict came the dramatic reduction of the US military. That meant many military installations were slated for closure. Base Realignment and Closure, or BRAC, was the dreaded acronym no community wanted to hear, especially those that depended on a nearby military base for economic support. In part, it was the political, militaristic, and economic upheavals in Europe that were the impetus for the changing military threats to the Western world. The Soviet Union had enough problems at home without sending out their submarine force and surface fleets as in the heyday of the Cold War.

For Naval Air Station Barbers Point, the writing was on the wall. After many hearings and debates, its fate was decided with the stroke of pen. It was slated for closure on July 2, 1999. After 57 years of proud service, the last naval air station in Hawaii would cease operations. From then on, it would be renamed Kalaeloa Airport.

Pictured here are the control tower and operations building in the early 1990s, when it was decided to turn the air station over to civilian control. It was essential to have it as an alternate airport for commercial carriers in case of emergencies. (LOC.)

This is the administration building on Lexington Avenue, across the flight line. After the closure of the base in 1999, this building sat shuttered until 2007, when it became American Renaissance Academy, a private school. (LOC.)

Aside from the P-3 Orions stationed there, one of the more familiar sights were the A-4 Skyhawks of Composite Squadron One (VC-1) Blue Aliis. This photograph is dated August 10, 1976, and shows the executive officer's airplane with a multicolored rainbow fin flash. On the canopy rail is the name of the executive officer, Cdr. Gerry Coffee, who was shot down in Vietnam and spent seven years as a prisoner of war in the infamous Hanoi Hilton. Coffee now resides in Hawaii and is very active in the community. (USN.)

This is an in-flight underside shot of the same aircraft. VC-1 also operated the single-seat version known as the Echo model or A-4E. This photograph is for the benefit of the modeler, as there are not too many photographs of the type from this angle. (USN.)

Two A-4s join up on a VP-4 P-3B and fly formation. Pilots could not get enough flying in these conditions—ceiling and visibility unlimited. (USN.)

Two TA-4Js form up on two Hawaii Air National Guard McDonnell Douglas F-4C Phantom IIs. One of the tasks VC-1 had was providing air combat maneuvering training for locally based or visiting fighter units. From the mid-1980s until the mid-1990s, Kaneohe MCAS fielded three F/A-18 Hornet squadrons. VC-1 had the role of playing the aggressor air force. (USN.)

Pacific Aviation Museum director Ken DeHoff (left) stands next to Naval Air Museum Barbers Point director Bradley Kaai Hayes and the final flight crew from VP-47 Golden Swordsmen from Kaneohe MCBH. In the last few months of its flying life, VP-9 gave the P-3C Orion in the background to VP-47 to use solely for landing and take-off practice. While the paint scheme reflects VP-6, which was disestablished years ago, VP-9 painted its squadron number on the nose. (NAMBP.)

The 160770 now sits chocked and chained as a quiet reminder of the heady days of the Cold War, and the legacy and history of this landmark air station. Now Kalaeloa Airport, it has returned to its original place name from before Captain Barber made his ill-fated voyage. But for those who lived and served there, and to many people in Hawaii, it will always be known as Barbers Point Naval Air Station. (NAMBP.)